Textile Progress

December 2010
Vol 42   No 4

The measurement of wool fibre properties and their effect on worsted processing performance and product quality. Part 1: The objective measurement of wool fibre properties

Anton F. Botha
and Lawrance Hunter

The Textile Institute

Routledge
Taylor & Francis Group
LONDON AND NEW YORK

First published 2010 by The Textile Institute

2 Park Square, Milton Park, Abingdon, Oxfordshire OX14 4RN
52 Vanderbilt Avenue, New York, NY 10017

Routledge is an imprint of the Taylor & Francis Group, an informa business

First issued in paperback 2019

Back Issues: Taylor & Francis retains a three year back issue stock of journals.

ISBN 13: 978-1-138-45873-4 (hbk)
ISBN 13: 978-0-415-66877-4 (pbk)

CONTENTS

Textile Progress
Vol. 42, No. 4, December 2010, 227–339

The measurement of wool fibre properties and their effect on worsted processing performance and product quality.
Part 1: The objective measurement of wool fibre properties

Anton F. Botha[a]* and Lawrance Hunter[a,b]

[a]CSIR Materials Science and Manufacturing, Fibres and Textiles Industrial Support Centre, PO Box 1124, Port Elizabeth 6000, South Africa; [b]Department of Textile Science, Faculty of Science, Nelson Mandela Metropolitan University, PO Box 77000, Port Elizabeth 6031, South Africa

(Received 2 March 2010; final version received 16 April 2010)

The world has moved away from subjective appraisal of raw wool characteristics and has entered an era of objective measurement and specification, and the raw wool trade is rapidly moving towards sale by total description which necessitates the accurate, rapid and cost effective measurement of all the raw wool characteristics important in determining price, textile performance and end-use. The development and availability of new technologies and equipment have enabled the objective measurement of many more raw wool characteristics than was possible in the past.

Over the past few decades, a considerable amount of research has been carried out worldwide on the effect of the raw wool characteristics on topmaking and spinning performance, as well as on yarn properties. This was done in order to gain a better understanding of, and to quantify, the effects of fibre and processing parameters on processing behaviour and performance and on the properties of the top and yarn and even the fabric. An important aim of the research was to improve the processing of wool and the productivity and cost effectiveness of the various processing stages. This research led to a better understanding of which raw wool characteristics influence textile processing behaviour and performance, as well as the product quality and end-use performance, and ultimately the raw wool price. On the basis of this, technologies and instruments were developed and commercialised for measuring the key raw wool characteristics rapidly, accurately and cost effectively. In parallel to this, the associated test methods were developed and standardised largely under the umbrella of the IWTO, many of these being adopted and used in raw wool marketing and trading worldwide.

This review covers the research and development carried out over more than half a century on the development and standardisation of technologies, instruments and test methods for the measurement of those characteristics determining the price and textile quality of raw wool and which are therefore important in terms of the global marketing and trading of raw wool. Research and development in this field is still continuing, but at a much lower intensity and pace than during the second half of the previous century.

Keywords: wool objective measurement; raw wool properties; wool base; vegetable matter; wool yield; fibre diameter; coarse edge; fibre length; staple length; Hauteur; fibre crimp; staple crimp; resistance to compression; fibre curvature; staple strength; wool colour; coloured fibres; dark fibres; medullated fibres; pigmented fibres; wool style

*Corresponding author. Email: afbotha@csir.co.za

ISSN 0040-5167 print/ISSN 1754-2278 online
© 2010 The Textile Institute
DOI: 10.1080/00405167.2010.486932
http://www.informaworld.com

1. Introduction

Although wool only accounts for some 2% of global fibre consumption, it remains a highly valued and sought-after fibre because of its outstanding comfort and aesthetic qualities, particularly in the top or luxury end of the market. Of the global production of some 1.2×10^9 kg of wool, approximately 56% is used in apparel, 42% in household (interior) textiles and 2% in industrial applications. Close on 80% of the Merino (fine apparel)-type wools are processed on the worsted system of producing wool tops. Fibre properties, particularly fibre diameter, play a major role in determining worsted processing performance (Table 1 [1]), the value of the product (Table 2 [2]) and the auction price (Figure 1 [3]) for such wools. There is also a close relationship between raw wool price and top price [4,5], according to Bell [6], discussing the various cost elements of a top. Fibre diameter alone can even account for as much as 70 to 80% of price variation of Australian (i.e. Merino-type) wool on a clean basis [7–11]. Nevertheless, it is important to bear in mind that the relative importance and contribution of the different greasy wool characteristics to price can vary significantly from season to season and from year to year and also for different wool types [12–14]. Graham [15] has discussed the effect on wool price, of changing various wool characteristics, such as staple length, including providing the sheep with coats (rugging), tabulating the length discounts applicable to 100-mm super fine hogget wools. Snowder [16] has reviewed the factors affecting the price of US wool. Rogan [9] has discussed the influence of various wool quality attributes on the auction price (clean basis) of Australian wool [17], concluding that, for example, for the first quarter of 1995, the mean fibre diameter accounted for 76.5% of the variation in price, staple strength for 8.2%, staple length for 2.2%, vegetable matter (VM) for 2.3%, colour for 1.5% and style for 1.3%.

Table 3 ranks the desirable parameters and methods for specifying Australian (Merino-type) and New Zealand (cross-bred-type) raw wool [18].

Ideally, the auction price of wool should accurately reflect demand and 'textile quality and value', but this is not necessarily the case. For example, Lamb et al. [19] challenged certain purchasing specifications placed by spinning mills on greasy wool, such as the price penalty on very long wools, as they impact on grower profitability. For example, in terms of price paid, there is an optimum staple length of around 85 mm for a 16-μm wool [20], increasing to 95 mm for a 23-μm wool [20], with a penalty for staple lengths longer than these [20], which is generally not justifiable in terms of processing performance and product quality [21,22]. Later studies [23] indicated a levelling off rather than a decrease, except for superfine wool with a high staple strength. Scrivener et al. [24] discussed the valuation of superfine wools in relation to top-making performance, concluding that in terms of their processing performance visually assessed more stylish wools did not justify their price premiums and vice versa. Drummond [25] emphasised that to ensure that the raw material reflects the requirements of a mill, it is of paramount importance that the mill develops wool purchase specifications on the basis of full objective measurement.

The wool pipeline, from grower to consumer, is a long, complex and costly one (Figure 2) [2], with very many factors playing a role in determining the processing route, conditions and performance, as well as the processing and manufacturing costs, product quality and cost and end-use applications. Not least of these factors are the raw wool fibre properties, notably diameter [26], which play a major role and can vary dramatically according to genetic factors (including breed), farm management practices (e.g. time of shearing [27]) and the environmental factors, such as nutrition [28], climate, disease and lambing, which prevail during the growth of the fibre. In addition to these factors, on-farm classing [29] and clip preparation, blending and mixing of wool, packaging and

Table 1. Relative importance of raw wool characteristics on worsted processing performance [1]. Reprinted from Anonymous, *Staple Measurements*, AWTA Ltd. Newsletter, Victoria, Australia, October 2007, with permission of AWTA Ltd.

Raw wool characteristic	Importance
Yield	****
Fibre diameter	****
Vegetable matter	****
Length	***
Strength/position of break	***
Colour	***
Coloured fibres	***
Fibre diameter variability	**
Length variability	**
Degree of cottedness	**
Crimp/resistance to compression	**
Staple tip	*
Age/breed/category	*
Style/character/handle	*

Notes: ****Most important.
***Majo.
**Secondary.
*Minor.

Table 2. Comparative influence of raw wool characteristics on value of product [2]. Reprinted from K.J. Whiteley, Wool Technol. Sheep Breed. 36(2) (1987) pp. 109–113, with permission of the International Wool Textile Organisation (IWTO).

	Scoured wool	Top/noil	Yarns	Cloth
Yield	XXXX	–	–	–
Vegetable matter	XX	XX	XX	E
Fibre diameter	XXXX	XXXX	XXXX	XXXX
Diameter variability	E	E	E	E
Staple length	XX	XXX	X	–
Length variability	E	E	–	–
Staple strength	XX	X	–	–
Strength variability	E	–	–	–
Crimp	X	X	X	X
Cotts	E	–	–	–
Staple tip	E	–	–	–
Colour	X	X	X	X
Dark fibre	E	E	E	E

Notes: XXXX: Highly significant.
E: May not be present, or significant only when limits exceeded.
X: Significant.

Table 3.   Desirable parameters for specific raw wool and test methods [18]. Reprinted from P. Baxter, Wool Technol. Sheep Breed. 44(1) (1996) pp. 29–38, with permission of IWTO.

| Property | Priority of rank assigned by research workers | | | Test method |
| | Australia | New Zealand | | |
		Woollen	Semi-worsted	
Mean fibre diameter	****	**	***	IWTO-28
Yield	****			IWTO-19
Vegetable matter	***	*****	******	IWTO-19
Staple length	***			IWTO-30
Staple strength	***			IWTO-30
Length after carding		***	*****	NZS-8719
Colour – yellowness	**	*****	*****	IWTO-E14
Colour – brightness	**	***	*****	IWTO-E14
Dark fibres	*	*	**	
Style	*			
Handle				
Resistance to compression				AS-3535
Bulk		***	***	NZS-8716
Short fibre (<40 mm)		*	**	
Medullation		**	**	IWTO-8
Kemp		*	*	

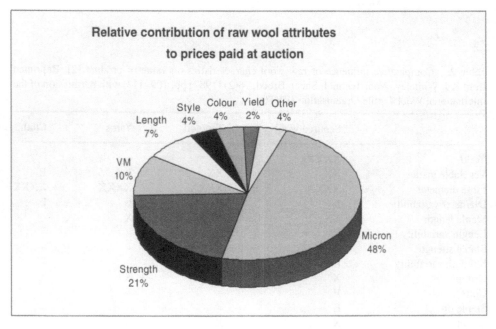

Figure 1.   Relative contribution of raw wool attributes to prices paid at auction [3]. Reprinted from R.C. Couchman, P.J. Hanson, K.J. Stott and C. Vlastuin, *Wool Quality: implications for worsted processing, grower receipts and R&D*, Workshop on Management for Wool Quality in Mediterranean Environments, Western Australian Department of Agriculture, Perth, Australia, 1992, with permission of AWTA Ltd.

storage [30–32] on the farm and during the early stage of processing, although not affecting the intrinsic wool fibre characteristics as such, can all affect the overall characteristics and uniformity of a bale or consignment of wool and which, therefore, can impact on processing performance and product quality. Various papers [7,29,33–83] have reported on and discussed issues related to grower and bulk classing, clip preparation, interlotting (lot building), covering aspects such as the associated implications and benefits, lot matching [79,80] and their effect on variability of staple length and fibre diameter and processing [43,44,49].

The Australian Wool Testing Authority (AWTA) introduced a wool uniformity index (UI) [84], which incorporates the coefficient of variation in diameter (CV_D), staple length (CV_{SL}), staple strength (CV_{SL}) and position of break to distinguish grower lots from bulk classed lots [85–88]. No differences between such lots varying in UI were observed [86,87, 89] in terms of processing performance and top quality. Hansford [79,90] described the use of objective testing of bales in store for lot matching, in terms of a new lot building system called Laser Matched Interlots (LMI), which uses both objectively measured mean fibre diameter and other subjectively assessed, commercially important parameters. Rottenbury et al. [53] concluded that the factors that dictate wool growth over a 12-month period, namely breeding, seasonal conditions and sheep management, effectively set the characteristics of a clip, and there is very little scope for wool preparation to significantly improve the processing behaviour of fleece wool apart from the removal of stained wool and VM. Moreover, non-fleece wool (e.g. skirtings) in blends with fleece wool does not generally affect predicted Hauteur, but does increase dark fibre risk [91]. Bazeley et al. [92] briefly summarised the various aspects related to clip preparation, including the Code of Practice for the Preparation of Australian Wool Clips, emphasising the overriding importance of staffing and organisation of labour in its successful implementation.

The overriding conclusion which can be drawn is that there is a close relationship between differences in processing parameters and the underlying changes in, and uniformity of, the raw wool characteristics of batches arising from preparation procedures, such as the

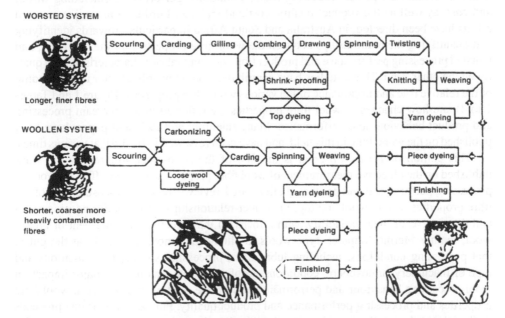

Figure 2. The steps involved in both woollen and worsted processing [2]. Reprinted from K.J. Whiteley, Wool Technol. Sheep Breed. 36(2) (1987) pp. 109–113, with permission of IWTO.

separation of inferior fleece components, and skirting and classing (i.e. in itself, preparation did not have a synergistic benefit on processing).

In 1983 Dittrich et al. [93] discussed developments relating to measuring techniques for testing wool quality, while Ponzoni [94] discussed the role of objective measurement in farm management, and Schaare and Jordan [95] discussed the impact of measurement technologies on animal production. Rogan [96] listed the following traits (for breeding objectives), in order of relative importance, which reflected wool value to consumers and processors; diameter, staple strength (using diameter variability), staple length, crimp definition, diameter variability, weathering and dust penetration. Cottle and Bowman [97] discussed the on-farm factors that affect greasy wool characteristics of importance to raw wool processors and ways in which the wool producers can meet the requirements and specifications of processors, while Rottenbury et al. [98,99] discussed mill consignments and the impact of sale-lot characteristics. Various publications [2,97,100–111] have discussed the requirements of the wool processor and manufacturer and also how the wool producer can meet the needs of the processor. Bell [112] discussed top making within the context of wool specifications. McMahon [113] discussed the various technologies involved in wool improvement, while Lee [114] gave valuable data and information relating to quality control aspects of worsted processing and Carnaby and Maddever [115] described a computer programme for formulating wool blends using linear programming.

In the light of the importance of greasy wool characteristics on price and performance, it is hardly surprising that for almost a century research and educational institutions, as well as industry, have devoted considerable time and effort in an attempt to objectively and accurately measure the various raw wool fibre properties. Major studies have been undertaken and very many papers have been published in this respect, notably in Australia, New Zealand and South Africa. The initial focus was on developing and standardising test methods (largely under the International Wool Textile Organisation (IWTO) umbrella), which can accurately, rapidly and cost-effectively measure those raw wool fibre characteristics which determine the price, processing route, conditions, performance (including waste) and cost, as well as the product quality, cost and end-use. Furthermore, major research efforts have been directed, in Australia and South Africa in particular, towards identifying and quantifying the relationships between the wool fibre properties on the one hand, and worsted processing performance and product (top, yarn and fabric) characteristics and quality on the other hand. These studies have contributed much towards a better understanding of the relative role and importance of the various wool fibre properties (Figure 1 and Tables 1 to 3) and how changes in wool fibre properties are reflected in downstream processing and product behaviour and performance. Many scientific and technical papers have been published on the work done in this field, and have been reviewed [116–120] at various times.

This edition of *Textile Progress* is the first of the series that will review the work published on the objective measurement of wool fibre properties and their effect on worsted processing performance and product quality. Part 1 reviews the actual measurement of the fibre properties of raw wool and top, their inter-relationships, changes during processing and the sources of their variability, with a focus on the objective measurement of greasy wool, mainly Merino 'apparel'-type wools. It should be emphasised right at the outset that processing conditions, including lubrication, processing speeds, the conditions and type of machinery, atmospheric conditions, wool regain etc. can have a major impact on wool processing behaviour and performance and on the relationships between wool fibre properties and processing performance and product quality. Furthermore, as one proceeds further down the wool-processing pipeline, the less impact certain raw wool properties have on processing performance and product quality. A good example is staple strength and position of break, which will affect the top fibre length characteristics, but have no other effect on spinning performance and yarn properties, except in those isolated cases when they provide a measure of intrinsic fibre strength.

Many publications have dealt with worsted processing conditions and technologies, and related aspects, and their influence on processing behaviour and performance, including factors such as entanglement during scouring, fibre breakage during carding, nep formation and removal etc. Nevertheless, these topics fall outside the scope of this review and will not be covered. The reader is referred to some good reviews covering these topics [2,30, 121–129].

2. The objective measurement of wool fibre properties

2.1. Introduction

2.1.1. General

The historical development, introduction and importance of the objective measurement of greasy and raw wool and the various associated tests and related aspects have been discussed in a number of papers [1,18,130–138]. A great deal of work was carried out, particularly in Australia, to introduce objective measurement and raw wool specification on the basis of pre-sale testing and sale-by-sample (sale-by-description) instead of the traditional subjective assessment, the introduction, development and marketing implications being discussed in various papers [134,139–166]. Whiteley and Rottenbury [164], while reviewing the research carried out into the specifications and marketing of Australian greasy wool prior to 1990, cited some 41 references, while details of the AWTA testing services are available at www.awta.com.au (testing the wool clip [167]).

In the 1940s and 1950s, the airflow and projection microscope instruments were developed to measure wool fineness, while the 1960s saw the development of greasy wool standard test methods for yield, VM and mean fibre diameter, the IWTO being pivotal in this respect. The 1970s ushered in the world of objective measurement [18], as sale-by-sample was introduced in Australia in 1972 [139,168] and, together with pre-sale testing, became part of the Australian wool-selling system in July 1972. This also happened in South Africa around the same time. Virtually 100% of the Australian wool clip was pre-sale tested by the early 1980s [168]. Standard sampling, conditioning, sample preparation, calibration and testing procedures and methods, such as those of the IWTO test methods and international inter-laboratory round trials (Interwoollabs [169]), are crucial for accurate and reproducible results, so important for trading purposes.

Between 1980 and 1985, prototype automated instruments for conducting additional measurements (staple length and strength) were developed by the Commonwealth Scientific and Industrial Research Organisation (CSIRO) and the South African Wool & Textile Research Institute (SAWTRI). The CSIRO made its instrument available to the AWTA Ltd for commercial testing, with additional measurement of wool staple colour and staple length and strength being introduced in 1985/1986 in Australia [9,168,170]. Similar systems were introduced in South Africa around the same time. The final report of the Australian Wool Board's Objective Measurement Policy Committee published in 1992 [171] spelt out details and ramifications of objective measurement of wool in Australia.

Baxter [131] and Sommerville [132] reviewed aspects of wool metrology as in 2002, discussing the recent developments and future directions. Marler [172] reviewed the historical development and industry reactions to colour and staple measurements over the preceding 20 years. Couchman [173] produced a summary report on the Australian Staple

Measurement Adoption Programme (ASMAP). Gleeson et al. [174] discussed the benefits
of staple measurement to the woolgrowers, processors and buyers and exporters.

Wood et al. [175] have discussed the objective measurement of New Zealand carpet
wools, covering properties such as colour, fibre diameter bulk, medullation, VM content and
length after carding. Wood [176] outlined commercial technologies for the testing of New
Zealand wool, certified testing of scoured wool having commenced in 1950. A programme to
develop a system of sale-by-sample was embarked upon by the Wool Research Organisation
of New Zealand (WRONZ) in 1967, such a system being introduced later in New Zealand.
Mahar [177] discussed the role of objective specification for carding wools that make up
approximately 15% of the Australian wool clip. Li et al. [178] discussed the introduction
of objective measurement in the Chinese domestic wool chain.

2.1.2. Fibre property distributions

As already mentioned, the various raw wool fibre properties largely determine worsted
processing performance, product quality and end-use application, as well as raw wool price
in the case of apparel-type wools. The accurate measurement of the wool fibre properties
is therefore critically important, from a technical, commercial and certification point of
view. A wool sample, whether it represents a staple, fleece, wool bale, farm sale lot, top-
making consignment, processed top, yarn or even fabric, contains fibres with a range of
diameters, lengths and strengths. The individual fleeces of any breed of sheep are composed
of millions of fibres that vary in diameter and length, there being something like 60 million
fibres in a single fleece. It is important to note that due to the inherent variability in wool
fibre characteristics within staples, between staples, between different positions on the
body of a sheep, between fleeces, within bales and between bales within sale lots, proper
sampling to obtain a representative sample for testing is of prime importance [179], with
core sampling [180–185] generally most appropriate for this purpose.

Due to the inherent variability in wool fibre properties, notably diameter and length,
appropriate sampling procedures have been developed and various mathematical models,
for example Gaussian (normal), log-normal, Poisson, gamma and their combinations etc.,
developed for describing populations with varying characteristics, have been applied to
wool [186–190]. The diameter distributions of fibres in raw wool and tops generally follow
a lognormal distribution [191–197]. Nevertheless, the most widely used model in the case
of wool fibres, notably for wool fibre diameter, is still the Gaussian or normally distributed
model, the mean and the standard deviation (SD) (or more commonly, the coefficient
of variation) being used to characterise the bell-shaped diameter distribution. Ford [198]
summarised work on the statistical properties of wool diameter and length, and discussed
a composite two-dimensional diameter–length distribution and its application to wool. In
reviewing the distribution of fibre length in commercial wool tops, Fish [199] noted that the
fibre length distribution in wool consignments had in recent years shifted from a bimodal
to a more even distribution, providing likely reasons for this.

In addition to measuring dispersion, it is sometimes also desirable to measure the extent
to which a distribution departs from normality (symmetry) in order to form an opinion
about the representativeness of various descriptive measures, the coefficient of skewness
being one way of doing so; whereas the normal curve is a precisely symmetrical bell-shaped
curve, which is dependent only on scale, a skew-curve is similar, but not symmetrical. It
has an extension to one side or the other. Relative to the normal curve it has too large
a proportion with either high or low values and is said to have a positive or negative
tail (skewness) accordingly. In addition to skewness, distribution curves can also deviate

from normality in terms of the sharpness (peakiness) or flatness of the distribution, but still retaining symmetry, this being referred to as kurtosis. If the major portion of the distribution is close to the mean and the minor portions extend well out on either side, i.e. sharp peak or long tails, this distribution has significant kurtosis and is named a leptokurtic distribution. The reverse of this, namely a flatter top with sharp shoulders and shorter tails, is called a platykurtic distribution. A measure of skewness is obtained from the second and third moments, while kurtosis is assessed by using the second and fourth moments.

Mayo et al. [200] discussed the determination of fibre diameter distribution, showing that for 13 Merino studs sampled in New South Wales, it was generally positively skewed and strongly leptokurtic, with skewness and kurtosis highly correlated. Diameter distribution tends to be more peaked and skewed for fine wools, with that of coarser wools more symmetrical [201]. Bow and Hansford [179] discussed fibre diameter and its distribution, particularly within the context of the role of diameter distribution as a specification of wool. It has been shown [201] that most variables of diameter distribution, such as skewness, kurtosis and coarse edge (CE), can be derived from the mean fibre diameter (D) and the standard deviation (SD). Gee [202] has also investigated the way wool fibre property distributions deviate from the normal distribution, and the implications thereof on processing performance and yarn and fabric properties.

This Chapter discusses the various test methods applied to greasy and raw wool and wool tops, particularly those forming part of the 'objective measurement and additional measurement programmes' and largely developed under the umbrella of the IWTO (www.iwto.org), details of which can be found at www.awta.com.au; testing the wool clip [167].

2.2. Wool base, vegetable matter base and yield

A bale of greasy wool can comprise anything from about 30–50% non-wool components, such as grease, suint, dirt, VM and moisture. Ward [163] summarised the large variability in the components (constituents) of greasy merino wool, Australian wool typically containing 15% dirt, 15% grease, 5% suint (dissolved salts), 4% protein contaminants and 2% VM. Clearly, the amount of clean wool in a bale of greasy wool is of overriding commercial importance. Yield generally refers to the amount of clean (usable) wool fibre, at a standard moisture content (regain) that is expected to be produced when a batch of greasy wool is processed. Testing for yield has basically changed little over the years and is normally determined according to IWTO-19–02, which incorporates the determination of *wool base* and *VM base*, which are measured and certified according to IWTO-19–03 and from which various standardised yields may be calculated by means of formulae contained in IWTO-31–02 and the IWTO core test regulations. Calculation of combined test certificates for yield and mean fibre diameter of raw wool in consignments is covered under IWTO-31–02.

Wool base is defined as the oven-dry weight of the wool fibre, free from all water solubles, grease (solvent soluble), mineral and alkali-insoluble matter, the latter generally representing the vegetable (plant) matter. Wool base is expressed as a percentage of the weight of greasy wool. The test for wool base entails a standardised scouring procedure, followed by drying to a constant weight, then the determination of residual grease, dichloromethane (DCM)-extractable matter (traditionally by Soxhlet ethanol extraction (IWTO-10) and recently also by the near infrared reflectance (NIR) technology [176,203,204]), residual mineral matter (dirt and sand) determination (traditionally by ashing at 750°C or recently also by the NIR technology) and VM determination. Research has shown that residual ash [205–208] and grease on the scoured wool, for the IWTO-19 yield test, can be measured by NIR, achieving an accuracy of ±0.1% or better when determining alcohol-extractable scouring residues in yield testing (IWTO-19). Most commercial laboratories achieve highly consistent scoured residual grease levels of between 0.5% and 1.5% [18]. Based upon international round trial results involving the NIR measurement of residual grease on commercially scoured wool and slivers, Ranford et al. [209] recommended that the working draft be upgraded to an IWTO Draft Test Method for scoured wool and sliver. Ranford et al. [210] developed revised IWTO-10 test methods for Soxhlet extractable matter determination on scoured and carbonised wool and combed wool sliver (top), the precision of the former being 0.20%.

Various types of VM can be found in bales of wool, the diverse types affecting processing performance and product quality differently. VM is generally separated into the following three broad categories:

• Spiral burr, clover burr and other soft burrs.
• Seed (grass seed), shive (the broken fibre like particles of grass seed) and other small particles.
• Hard heads and twigs.

High levels of the first two categories of VM can significantly lower the top and noil yield.

Anson [211] has discussed the identification and separation of VM types in core samples, including the industry requirements and a comparison of subjective and objective methods. Various papers [212–214] have reported on the subjective and objective estimation of VM levels and type in core and scoured wool samples. *Vegetable matter base* is the oven dry weight of VM, namely weight of the ash-free and ethyl alcohol extractive-free burrs (including hard heads), twigs, seeds, leaves and grasses present, expressed as a percentage of the weight of the greasy core sample. This is also printed on test certificates, as the percentage of hard heads, namely ring burrs (Sida platycalyx), Noogoora burrs (Xanthium occidentale), Bathurst burrs (Xanthium spinosum) and similar burrs of a bean-like or woody character covered in readily removable spines (which do not contribute to a loss of wool during processing) [215] and twigs, namely the small pieces of stick, woody leaf stalks and similar woody material (which do not contribute to a loss of wool during processing) [215]. The VM is determined in terms of quantity and type (the latter generally subjectively classified, although the use of NIR spectroscopy and image analysis for this purpose has also been investigated) after dissolving the wool in hot caustic soda (sodium hydroxide) and dissecting the residue to determine the type of VM. The type of VM (e.g. seed, burr, hard heads and twigs), identified with the aid of IWTO photographs, is important, as different types respond differently to caustic soda (and therefore appropriate correction factors need to be applied). Furthermore, the different types of VM differ in terms of their commercial importance, affecting processing performance and product quality differently; some, such as shive and grasses, are difficult to remove during processing, while others are associated with greater fibre losses when they are removed during processing.

Cleanliness faults (VM etc.) can be measured visually in sliver and tops using balanced illumination (IWTO DTM-13) or automatically using the Centexbel Optalyser (IWTO-55).

2.3. Fibre diameter

2.3.1. Introduction

As mentioned previously, mean fibre diameter can account for as much as 80% of worsted spinning performance and variations in fabric handle, finer fibres generally being superior, except for fibre entanglement during scouring and fibre breakage and nep formation during mechanical processing. Diameter is also the limiting factor when it comes to spinning fine quality worsted yarns. It is therefore not surprising that, for commercial and trading purposes, as well as for technical reasons, considerable effort has been directed towards the accurate, rapid and cost-effective measurement of this important wool fibre characteristic, as well as the evaluation and standardisation of test methods and instruments. Rogan [216] and Lamb et al. [217] briefly reviewed the effects, from the farm to the finished fabric, and economic importance of diameter and diameter variation for Australian wool.

Roberts [218] has reviewed the factors that can affect the diameter of Merino wool, fibre diameter being dependent upon genetic factors (e.g. breed), as well as environmental (growth) factors, such as nutrition, climate, disease, insects, lambing etc. For example, fibre diameter and length are related to feed intake for a particular sheep, while stress caused by climatic conditions (e.g. extreme cold), disease, hunger, insects, lambing etc. decrease diameter, often causing a sudden and localised decrease (break or tenderness) in fibre diameter, in unison with the staple.

It is a well-established fact that wool fibres display considerable variation in diameter not only along their length (which can vary by up to 10 μm) but also between fibres within a staple (mainly affected by genetic factors [179]), between staples within a fleece [219–223] and between fleeces within a flock, each of which can affect processing performance and yarn and fabric properties in a different way. The CV_D within a fleece is around 20%, even under adequate conditions of nutrition [222]. The average range in fibre diameter (from minimum to maximum), along with the profile of staples sampled in 1997/1998 from Western Australian sale lots, was found to be 6.1 μm [224]. Wang [225] concluded that within-fibre (i.e. along fibre) diameter variations are common for wool and other animal fibres and have a large impact on the mechanical properties of the single fibres. It is worth noting that prior to around 2000, little quantitative distinction was made between along-fibre and between-fibre variability when referring to fibre diameter variability and distribution, results and research generally being based on measurements that combined the two sources of fibre diameter variations. At the outset, it is once again important to note that wool fibre diameter variation occurs in different forms and ways, essentially along and between fibres, and that the different sources of variation can impact differently on processing [219,226]. For example, the along-fibre variation is important because it determines not only the wool fibre strength but also the position where the staple, and more importantly the fibres, will break, which in turn significantly impacts the early stage processing performance and the fibre length and length distribution of the top. With the introduction of advanced and cost-effective technologies for measuring the different components of fibre diameter distribution, a more precise measurement and understanding of all the above variations in wool fibre diameter, especially within staples, has become possible. For example, using such advanced technologies, Baxter [227] could demonstrate the occurrence, though very rare, of as fine as 5-μm wool fibres.

As already mentioned, where nutrition, lambing, disease and other stress factors impact on a sheep, the fibres growing together tend to change diameter in unison, resulting in a section (segment) within the staple having an overall smaller cross-section, commonly referred to as a 'break' or 'tenderness'. In severe cases, such a 'break' is clearly visible, and if the staple is gripped at its two ends and extended, it inevitably breaks at such a position of minimum cross-section. Therefore, if a staple is chopped into a series of segments along its length and the mean fibre diameter of each segment is measured, the outcome is a picture of the average fibre diameter changes throughout the entire fibre growth period. Table 4 [228] gives an example of the relative sources of mean fibre diameter variation within a

Table 4.   Variation of fibre diameter within a mob [223,228]. Reprinted from B. Quinnell, K.J. Whiteley and E.M. Roberts, *Variation in fibre diameter of wool fibres—A review on objective measurement of wool in Australia*, Tech. Rep. of the Australian Wool Board's Objective Measurement Policy Committee, Australian Wool Corp., Melbourne, Australia, October 1973, pp. 4.2–4.20 and P.R. McMahon, "*Wool quality specifications in an integrated production and commerce manufacture system*", in *Proceedings of the 5th Int. Wool Text. Res. Conference*, 4, Aachen, 1975, pp. 1–7, with permission of The Textile Institute.

Source of variation	Percentage of total	
	Sound	Tender
Within staple between fibres	64	43
Within staple along fibres	16	43
Within fleece between sites	4	3
Between fleeces	16	11
	100	100

mob (i.e. a collection of sheep from one farm). These different components of variance are important when considering the most appropriate sampling technique required to produce a reliable measure of fibre diameter distribution for a fleece. Hansford [229,230] mentioned that an important requirement for measuring fibre diameter distribution is that the sampling procedure and sample are appropriate to the source of variation, and consequences thereof being investigated. It is also important to note that, for Merino sheep, wool fibres tend to grow with a constant length-to-diameter ratio, i.e. coarser fibres tend to be longer than finer fibres, also in a staple [120,179].

As can be seen from Table 4, the within-staple, between-fibre diameter variation is the largest for sound wool, but the picture changes in the case of tender wool, where the variation along the fibres (43%) contributes proportionally more to the overall variation. For sound wools, the between-fibre variation (i.e. excluding the within-staple along fibre component) accounts for over 80% of overall variation, whereas for tender wools it accounts for less than 60% [224]. Yu [231] found that, for Western Australian wool fleeces, the within-fibre diameter variation was as large as that between fibres in a staple.

A large number of papers [232–234] have discussed the fineness (diameter) testing of wool fibres and associated issues. An excellent series of highly informative and readable technical articles on the different principles and methods of measurement of wool diameter characteristics have appeared over the past decade in the AWTA Ltd Newsletter and the reader is referred to them for additional technical details and information (www.awta.com.au).

Due to the overriding importance of fibre diameter in determining wool price, processing route and performance and application, considerable effort has been directed towards its measurement, with many techniques and principles being developed and applied for this purpose, including the following (only certain references given below, others being quoted later):

- Airflow
- Sonic [235–238]
- Projection microscope [239]

- Radio isotopes [240]
- Liquid scintillation spectrometry [238,241,242]
- Lasers [243]
- Light diffraction [244–253]
- Image analysis [254–257]
- NIR [258]

Traditionally, mean fibre diameter was measured using the projection microscope (IWTO-8–04) and airflow methods. The constant pressure airflow technique was accepted by the IWTO for measuring wool sliver and tops in 1960 (IWTO-6) and for greasy wool in 1971 (IWTO-28) [259]. The airflow method is based on the Kozeny equation [260] covering the flow of a fluid through a porous medium having variable particle sizes and shapes, applied to wool fibre diameter (fineness) measurement early in the twentieth century [260], it being based upon a constant fibre density. The airflow method (IWTO-6 for combed sliver and IWTO-28 for raw wool core samples) was the de facto standard for the trading of greasy and scoured wool and wool tops from 1971–2000 [260], sample preparation being important in ensuring accurate results. Even today, and despite technical advances and new instruments being developed, fibre diameter is often measured by equipment functioning on the airflow principle because of the accuracy and cost effectiveness of this method. The airflow measurement of fibre diameter has been reviewed [260]: Sommerville [261] discussing the principles involved in the airflow measurement of fibre diameter, and Baxter [262] providing a good overview of airflow and projection microscope measurement of diameter and the factors influencing the results. Doak and Mahar [263] developed a technique for measuring mean fibre diameter by airflow method using a test specimen that can vary in mass (from 2.4 to 2.6 g).

The main weakness of the airflow method is that it only measures the mean fibre diameter, which is no longer regarded as sufficient information, because the diameter distribution also affects textile performance and wear comfort. It also has some relatively minor disadvantages in terms of potential effects of the CV_D [261,264–275], trash, medullated fibres (fibre density) and lambswool [266] on the results. In 1949 Anderson and Warburton [276] demonstrated the dependence of airflow method-measured mean fibre diameter (D) on the fractional coefficient of variation (C) of fibre diameter, viz. $D = D_1 (1 + C^2)$, where D_1 is the mean fibre diameter measurement by the projection microscope. Clearly, the CV_D of the sample being tested relative to that used for calibration is the deciding factor. Various other papers [272,277,278] have also dealt with the effect of CV_D on airflow measurement results. As an extreme example, if the components of a blend differ by more than 10 μm in mean fibre diameter, the difference between airflow method-measured mean fibre diameter for the blend can differ by as much as 4 μm from that measured by other instruments, such as Laserscan and Optical Fibre Diameter Analyser (OFDA), due to the effect of CV_D on the airflow measurement readings [265], this being in accordance with theoretical considerations [264,265,271,276,279]. Under the conditions normally encountered in practice, however, the effect of CV_D is generally small, mostly insignificant [265,280] (a 3% higher than 'average' CV_D increasing airflow method-measured mean fibre diameter by 0.2 μm [271]). The effect of medullation on airflow method-measured diameter is due to its effect on fibre density, a difference of 0.013 g/cc in density leading to a difference of 0.29 μm in mean fibre diameter [260].

The projection microscope method has disadvantages, as it is labour-intensive, length biased, very time consuming and can also exhibit an operator effect. Nevertheless, even today it remains the reference test method [281]. The effect of snippet length on projection microscope diameter measurements was investigated [282–285], the effect increasing with mean fibre diameter, but being constant for snippet lengths between about 0.6 mm and 1.4 mm, the effect being related to fibre ellipticity, which increases with increasing diameter [286]. According to inter-laboratory trials, the 2 mm snippet length produced slightly lower mean fibre diameter and SD values [283,284]. The effect of fibre non-circularity (ellipticity) on profile methods of measuring fibre diameter has been investigated by various workers [283,285,287–294].

In the light of the above, a number of improved methods, of measuring both mean fibre diameter and distribution of diameter, such as by laser and image analysis, have been developed due to their advantages in terms of speed, accuracy and cost-effectiveness and the fact that they simultaneously measure fibre diameter distribution, and today, even fibre curvature. Examples of such automatic single fibre diameter measurement instruments include the Sirolan-Laserscan™ and the OFDA100, which have been developed to provide an accurate measure of both the mean fibre diameter as well as the fibre diameter distribution and which have been accepted for use in the objective measurement of wool fibre diameter and its certification for trading purposes.

The use of photometry (analytical use of light intensity) to measure fibre diameter has been reviewed [295,296].

2.3.2. Laser-based instruments

The application of laser technology for the measurement of single fibre diameter was reported some 40 years ago [295–297]. One of the first applications of laser optics to fibre diameter measurement was by Lynch and Thomas in 1971 and 1976 [295,296], involving the transport of fibre snippets carried in a suitable fluid (e.g. isoproponal/water mixture) past a laser beam, constructed by the CSIRO and known as the Fibre Fineness Distribution Analyser (FFDA).

The basic principles of the laser-based Fibre Diameter Analyser (FDA) were described in 1976 [298,299], with David [300] reporting on its precision. The FDA was based upon an electro-optical measurement of the amount of light scattered from a directed beam by fibre snippets, which are dispersed in a liquid and transported past the light beam in the moving liquid. Various papers [301,302] have dealt with the FFDA and the FDA.

Due to various problems experienced with earlier laser-based instruments, such as the FFDA and the FDA [188,304–307], notably their optical discrimination system resulting in biases and non-fibrous fragments and particles also leading to errors, the CSIRO decided to develop a new and completely redesigned instrument which is now marketed as the Sirolan-Laserscan™ (Figure 3). The development, performance and application of the Sirolan-Laserscan™ , which has been in commercial use since 1992, have been reviewed by Charlton [308]. This instrument incorporates the latest technology, particularly in terms of its optics and electronics, and is widely used today, particularly when it comes to objective measurement of greasy wool for certification and trading purposes. A schematic diagram of the Sirolan-Laserscan™ system is shown in Figure 4 [218]. A small sample (approximately 0.03 g) of 2 mm-long fibre snippets obtained by a mini-coring procedure for greasy wool and by means of a guillotine for top or sliver, is dropped into a dispersing bowl filled with 92% isopropanol and 8% water. The snippets are dispersed and gradually carried down through a measurement cell, where they intercept a laser beam. The transmitted signal is split into two parts. Part of the signal is used for the measurement, with variations in the beam intensity being converted directly into fibre diameter. The other part of the signal is analysed by a patented fibre-optic discriminator [218], which enables the system to reject

Figure 3. The Sirolan-Laserscan™ [303]. Reprinted from AWTA Ltd. Newsletter, 50th Anniversary Edition, October 2007, with permission of AWTA Ltd.

invalid measurements caused by dirt, VM, fibre fragments, multiple and crossed fibres and fibres only partially intersecting the beam [308].

The fibre-optic discriminator only accepts single, intact fibre snippets. After passing through the cell, the sample and fluid are filtered through a cleanable primary sump filter before the fluid is recirculated through a replaceable secondary filter back into the glass dispersion bowl. Thus, a snippet can be counted only once. The optical and fluid systems are maintained at a constant temperature of 20°C for stability and for measurement at standard conditions [218].

Various papers have been published on the Sirolan-Laserscan™ system [259,309–312], with proving and calibration trials having been reported to the IWTO in 1992 [259,309, 313,314], an international round trial in 1993 [310] and Sommerville [283] reporting on the introduction of the Sirolan-Laserscan™ by the AWTA Ltd for the certification of mean fibre diameter.

After extensive international inter-laboratory trials on raw wool and wool tops, the draft test method for the Sirolan-Laserscan™ was advanced to a full test specification [315–317], the test method being ratified, covering its operation as a full IWTO test method, at the 1993

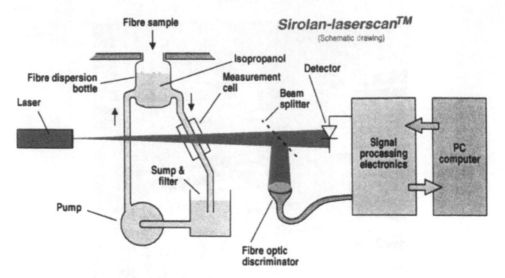

Figure 4. Schematic diagram of the Sirolan-Laserscan™ system [218]. Reprinted from E.M. Roberts, Wool Technol. Sheep Breed. 17(2) (1970) pp. 27–30.

Nice, France meeting of the IWTO [317]. In 1995, the IWTO approved, for certification purposes, the new specification (IWTO-12–95) for the determination of mean fibre diameter of raw wool and wool sliver [280], based on the Sirolan-Laserscan™ , with Marler [318] proposing an amendment to include Range Checks in IWTO-12. In June 2007 the Sirolan-Laserscan™ was chosen as a certified IWTO method (IWTO-12–03) for measuring fibre diameter distribution.

Treloar [319] concluded that woollen-spun yarns can be tested for fibre diameter on the Sirolan-Laserscan™ , provided acceptable dispersion of the fibre snippets is achieved during the washing of cut yarn sections, supporting earlier recommendations [320] concerning amendments to IWTO-12–95 to describe procedures for testing woollen and worsted spun yarns for fibre diameter distribution. Holcombe and Papazian [321] described a technique for samples and preparation of snippets from woven and knitted fabrics, using a prototype device for separating fibre snippets in yarns by an air separation technique.

2.3.3. Image analysis based instruments

Image analysis lends itself to the rapid and automatic measurement of wool fibre diameter, various instruments and test methods based upon the automatic measurement of projected fibre images having been developed over many decades. These include the Fibroalineator developed in the early 1950s [295,296], extended to the development of the Electronic Fibre Fineness Indicator (EFFI) in 1957 [297], which automatically scanned cut fibres mounted on a microscope, used on wool tops in 1970 [295,296], Fibre Diameter Video Analyser (FIDIVAN) [176,283,297], introduced in 1972 at the IWTO meeting in Monaco [297], PiMc, Fibre Image Display and Measurement instrument (FIDAM) [322–324] and OFDA, both the latter being based upon the original WRONZ FIDIVAN concept [176].

During the period the laser-based instruments were being developed, an instrument based on the image-analysis principle, called FIDAM [254,257,283,322–324], was being developed. The FIDAM involved proprietary software, a low power microscope and a

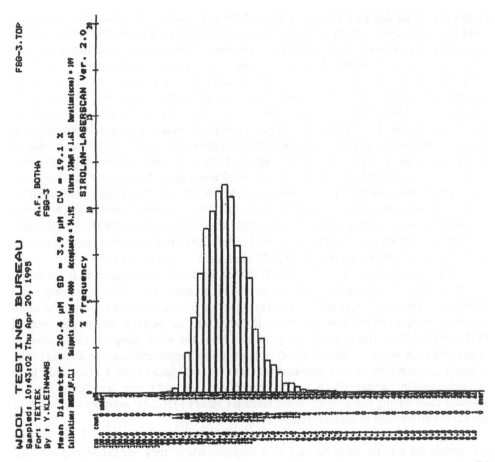

Figure 5. Example of fibre diameter distribution produced by means of the Sirolan-Laserscan™ system.

CCD camera interfaced to a PC, comparing well with airflow measurement results on raw wool samples [322–324], and was developed by Mark Brims at the AWTA Ltd [297] in 1981. After the AWTA had made the decision not to commercialise the FIDAM for raw-wool testing, Mark Brims (BSC Electronic Pty Ltd) in association with SGS Wool Testing Services (Australia) took advantage of advances in image analysis hardware and software and in 1989 designed and built an automatic image analyser to measure fibre diameter. The system, called the optical-based –fibre –diameter analyser (OFDA), uses similar physical components as the FIDAM (microscope, camera, computer etc.), but uses completely different methods of image acquisition and analysis [325–328]. Brims [224] reviewed the development of the range of OFDA instruments, from the development in 1990 [329] and release in 1991 of the original OFDA100, which is essentially an automatic microscope. Various papers have been published over the past two decades describing the measurement principles, applications, accuracy and precision of the OFDA [225,325–328], including round trials, the results being sensitive to snippet length, but less so at around 2 mm snippet length. Blankenburg et al. [330] suggested that a 2 mm snippet length should be recommended in IWTO-8, and that wools with high ellipticity should not be used as reference tops. Fibre relaxation has a small, though significant, effect on the

OFDA-measured diameter, this being allowed for in two different calibrations for greasy wool and wool sliver, which require different preparation techniques for the calibration tops [331]. Nevertheless, such calibrations for mean fibre diameter introduce a bias for CV_D, and vice versa, hence different calibrations are required when certifying each. In 1991, the IWTO adopted the IWTO-47–92 (E) method – measurement of the mean and distribution of fibre diameter of wool using an image analyser as a Test Method under Examination (TME). This test method broadly covered image analyser systems, such as the OFDA, not just the OFDA [255,332]. After comprehensive international round trials on raw wool and wool tops [315,316], the draft test method for the OFDA100 was advanced to a full test specification. The IWTO approved the new specification (IWTO-47—95) in 1995 for the determination and certification of mean fibre diameter for raw wool and wool sliver [280], and appropriate amendments to IWTO Core Test Regulations and Blue Book being accepted in December 1996 [280]. In 1998 it was established as an IWTO test method and today is an accepted IWTO certification method under IWTO-47–00 [333], with amendment approved in 2007 and issued in July 2007, following two round trials, and Couchman [334] discussing the OFDA2000 when used in the '100 mode'.

The physical layout of the OFDA100 system is depicted in Figure 6. It combines the properties of a projection microscope with computerised image processing, and measures 2 mm fibre snippets from combed sliver or from scoured raw wool. The rapid and easy sample preparation procedure is based on the popular microtome method IWTO-8–04 [335]. The snippets are produced from the wool samples, either using a mini-corer (raw wool or scoured wool) or a guillotine (applicable only to a sliver). The snippets are conditioned in a standard atmosphere and spread onto a glass slide by a snippet-spreading device which accepts snippets of 1.5 to 2.0 mm length, the type of spreader used having some effect on the diameter values [336,337] possibly as a result of associated differences in fibre curvature [336].

The glass slide is inserted into the slide preparer. Using a pair of tweezers, the snippets are introduced into the preparer as 2 to 5 aliquots, where they are distributed over the slide by a rotating spreader paddle. The spreading process takes less than a minute and is designed to evenly spread the snippets so that an obscuration ratio of snippet area to slide area of about 15–25% is achieved [255]. Once the snippets have been spread to the required density, the cover slide (the second glass slide) is hinged closed (fixed) and secured to the other and can be placed into the clamps on the microscope stage. Automatic scanning of the fibre sample captures the exact image of the snippet. The method recommends that the instrument setting remains constant and the whole slide is measured using the 'whole slide × 2' option to ensure that at least 4000 snippets are measured.

The OFDA100 software program is so designed that the 'intelligent selection process' prevents both the double measurement (measuring the same position that has been previously measured) and the measurement of superimposed fibres (two fibres clinging together). A fibre sample, involving the accepted measurement of 2000 to 5000 snippets takes only about 50 seconds to scan. When the measurement is completed, the resulting data, including mean fibre diameter, standard deviation, coefficient of variation of fibre diameter and prickle factor (PF), are displayed on the data screen or printed out on a histogram plot as shown in Figure 7. The monitoring screen, which comes with the OFDA100, continuously displays the magnified image of fibres during the measuring procedure. The displayed fibres and the measuring points can be captured in the form of a 'snapshot' at any time, saved and retrieved whenever required for subsequent checks or comparisons. All functions are program-controlled via the software. The OFDA100 also offers options for saving, sorting, displaying and printing the measurement data as well as the images shown on the monitoring screen.

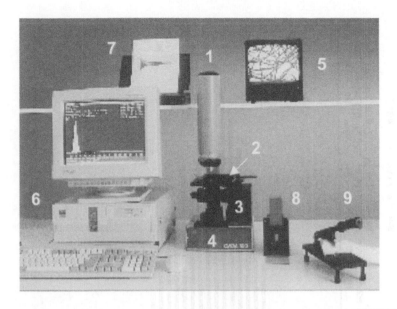

1 – CCD camera	6 – Computer with image acquisition
	hardware and software
2 – Traversing stage	
3 – Stage stepper motors	7 - Printer
4 – Illumination and interface unit	8 – Fibre spreader
5 – Video monitor	9 - Guillotine

Figure 6.   The physical layout of the OFDA100 instrument, reprinted with permission of Mark Brims.

Edmunds [337] gave some preliminary results illustrating the application and the potential of the OFDA for measuring additional fibre parameters, such as medullation [338] and fibre curvature [262,339,340]. Xu and Ting [341,342] have discussed the image analysis of geometrical properties of fibre, including the application of specially designed algorithms.

2.3.4. *Fibre diameter distribution characteristics*

2.3.4.1. Introduction. Several papers have dealt with the measurement and some of the implications of fibre diameter distribution [201,343–347], with Crook and co-workers [299,348] investigating the relationship between the fibre diameter variability and the wool quality.

Research carried out over many years has demonstrated the role of the fibre diameter distribution characteristics and variability on processing performance and product quality, including fabric and garment comfort. Fibre diameter variability, notably along-fibre variability, has a highly significant effect on fibre and staple strength and therefore on fibre

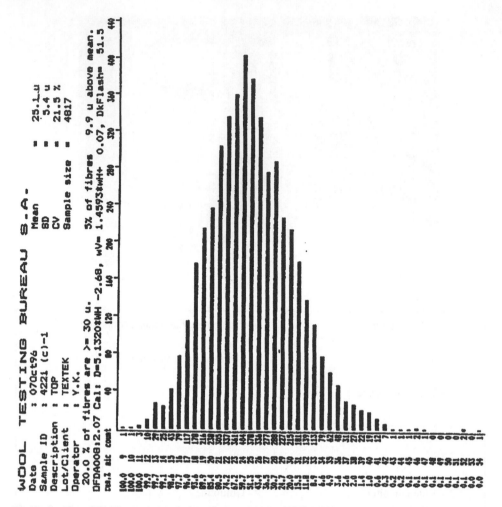

Figure 7.　Example of a histogram plot obtained on the OFDA100.

breakage during processing, notably carding, as the fibres mostly break at their thinnest places. Fibre diameter variation has a small, though significant, effect on the average number of fibres in the yarn cross-section, spinning performance, yarn evenness, thickness faults and neps. It is related to fabric handle and prickle (scratchiness), particularly as much as it is associated with an increase in the CE.

2.3.4.2. Fibre diameter distribution. From previous discussions it is clear that any two wool samples, having the same mean fibre diameter, could have widely different fibre diameter distribution characteristics (see Section 2.1.2), including the percentage of fibres having a diameter greater than a particular threshold value, commonly referred to as the CE.

2.3.4.3. Coarse edge. Considerable work (reviewed by Botha [118]) has been carried out to determine the origin of the problem of prickliness or scratchiness often associated with wool garments worn against the skin. It was shown that it is essentially the fibres coarser than about 30 μm, labelled as the coarse edge or CE, which were largely responsible for the sensation of prickliness or scratchiness, and was frequently ascribed incorrectly to an allergic reaction to wool. It was also shown that changes in CE of as little as one percentage point (absolute) can lead to differences in perceived prickliness. It is thus important for this parameter to be measured accurately, the more so because some farmers are now trying to include this information as part of their breeding criteria. Considerable work by the CSIRO in Australia has shown that once the CE exceeds about 5%, most wearers will experience discomfort (prickliness or scratchiness) in garments worn against the skin. Later work has indicated that it is the diameter of the fibre ends and not that of the fibre as a whole that is relevant in determining the CE of a sample, as the ends normally project from the fabric surface and come into contact with the skin. Naylor [349] developed a technique for the measurement of the diameter of the fibre ends in tops that was incorporated into the IWTO Draft Test Method (DTM-60 – 01), which also defines a fibre ends fineness index (FEFI) [350] that is related to the difference between the mean fibre diameter of the top and the mean diameter of the fibre ends [351].

For an accurate measurement of the CE, a sufficiently large number of fibres are to be measured for diameter [349,352]. Modern instruments, such as the Laserscan and the OFDA, lend themselves ideally for this purpose, at least 4000 fibres generally being measured. Naylor [352] used an approach, based on the definition of CE, to place the data into two groups, with all measurements being allocated to either the group labelled 'coarse edge' or to the group labelled 'not coarse edge'. By reducing the diameter distribution to only these two groups, the number of fibres in the 'coarse edge' group in any measurement of diameter distribution, should be readily described by binomial statistics, with the percentage error inversely proportional to $\sqrt{n}$, where n refers to the total number of counted fibres. For example, increasing the sample size by factor 10 would decrease the percentage error by factor $\sqrt{10}$. Thus, the precision of the number of fibres measured in the coarse tail is directly related to the total number of fibres measured for the sample and is a function of the appropriate statistics.

Botha [118] found similar mean fibre diameter and CE levels for different breeds of wool sheep in South Africa, although overcrimped wools (relative to Duerden standards) tended to have higher CE levels than 'average' or 'undercrimped' wools. He also found that 'inferior' wools had higher CV_D and CE levels than those of good quality wools of the same mean fibre diameter.

Considerable work was carried out to establish the relationship between CE and other measures of fibre diameter distribution, such as mean and SD, and this has been covered in Section 2.3.6.4.

2.3.5. General

Many studies [264,265,268–270,311,316,353–364] have been carried out to determine the relationship and agreement between the diameter measurements obtained on different instruments. Most of these studies have found highly acceptable agreement between measurements of different instruments. Some discrepancies were observed, however, particularly at the extreme ends of the diameter range; for example, between the Sirolan-Laserscan[TM] system and airflow method results at the superfine end of the scale [284,360], which was ascribed [360,365], at least in part, to an extrapolation error associated with the airflow calibration. Baxter [227] also explored the reasons for differences between the OFDA and the Sirolan-Laserscan[TM] system, in terms of the fine tail and the fibre diameter distribution shape for ultra-fine grower lots, concluding that fibres as fine as 5 μm, as measured by the OFDA, do actually exist. Heath et al. [366] found an overall difference of 0.15 μm between airflow and the Sirolan-Laserscan^{TM} measurements on South African wools, with no apparent effect of mean fibre diameter on the difference. Baxter [367] found that extreme degrees of medullation did not affect the agreement between the OFDA and the Sirolan-Laserscan^{TM} diameter results, but, as could be expected, did affect differences between their results and those of the airflow measurement, thereby suggesting that the airflow measurement results could be predicted from various OFDA-measured parameters. Marler and Harig [362] reported that differences between airflow method diameter values and those obtained by other instruments could be reduced, but not entirely eliminated, by correcting for CV_D.

Various workers [288,313,337,341,354,368–383] have studied the effects of various parameters, such as fibre snippet length and relaxation, on the Sirolan-Laserscan^{TM} and the OFDA diameter and curvature values. Both the fibre snippet length and the state of relaxation have some effect on the diameter values, with the state of relaxation affecting the diameter values most, and more particularly those of the OFDA [288,337,368,377,380]; hence, the use of different greasy wool and top calibrations for the OFDA [354]. Edmunds and Ranford [372] suggested that differences in mean fibre diameter obtained by different researchers, as well as those obtained on the OFDA because of relaxation, could be due to a complex interaction between, and changes in, snippet length and fibre curvature, the latter being affected by factors such as the state of relaxation of the fibres. Differences between the OFDA and the Sirolan-Laserscan^{TM} values due to sheep breed and country of origin [384] and 'ageing' of tops [385–387] could be due to the effect of fibre relaxation on diameter [377]. Moving away from the 2.0 mm sampling devices (guillotine and minicore) introduces a fibre snippet length effect, causing potential errors in the diameter values [337].

Baxter [262] reviewed and summarised the important measurement principles of the airflow and projected image methods, discussing differences that may arise due to various influences providing a detailed review of published work. Baxter [388,389] presented a summary of papers published during the previous 30 years on the effect of fibre relaxation on fibre diameter measurements, noting the discrepancies in results obtained and providing possible reasons for such variations. He summarised the effects as follows [388]:

- Airflow: Relaxation mostly increases mean fibre diameter as a result of an increase in fibre crimp and fibre-to-fibre contacts reducing the 'apparent' fibre surface area.
- OFDA (and possibly the projection microscope): Relaxation given to tops decreases the measured fibre diameter values, as a result crimp restoration producing a lower proportion of major to minor axis projections in the OFDA image (probably also that of the projection microscope).
- Projection microscope: Conflicting results have been published, effect possibly similar to that of the OFDA.
- Sirolan-Laserscan^{TM}: The effects, if any, of relaxation treatment not clearly documented.

Essentially, it is important to treat the wool samples to be tested in exactly the same way as the calibration samples were treated.

Knowles et al. [356] in 1998 reported significant differences between airflow and the Sirolan-Laserscan^{TM} diameter results over the entire range (16 to 43 μm) of New Zealand wools, with the airflow and the OFDA measurements showing better agreement within the range covered by the IH tops, the OFDA and the Sirolan-Laserscan^{TM} giving similar results of above 40 μm. The effects of CV_D were small over most of the diameter range, but were significant for lambswool and above 40 μm, the medullation affecting the airflow method results, but not those of the OFDA or the Sirolan-Laserscan™ results.

Wool fibres tend to be elliptical rather than perfectly circular, the effect of fibre ellipticity on fibre diameter measurements having been reported by various workers [283,285,287–294], with an average value of 1.08 for the maximum to minimum fibre diameter ratio (i.e. ellipticity) being fairly typical for Merino-type wool [287,390]. Blankenburg et al. [292–294] addressed the complication introduced by fibre ellipticity in fibre diameter measurements involving profile (micro-projection and the OFDA) techniques, and the role of snippet length, fibre crimp/curvature and relaxation in this respect. They hypothesised that a relaxation treatment brought the major/minor axis to a random position on the slide, which results in a diameter value closer to the 'apparent diameter', this being supported by Edmunds [391].

Brims [392] did a theoretical comparison of the OFDA and the FDA class instruments used for measuring fibre diameter distribution, comparing the effects of certain fibre parameters (i.e. curvature, blobs etc.) on such instruments with those on instruments such as the Sirolan-Laserscan™ and projection microscope. Baxter [262] reviewed and summarised the important measurement principles of the airflow and the projected image methods, discussing differences that may arise due to various influences, and providing a detailed review of published work in the process.

Sommerville [283,393] and Sommerville and Teasdale [280] reviewed the development of the projection microscope, the airflow, the Sirolan-Laserscan™ and the OFDA methods, including the precision and differences in their measurements, thereby giving the following table (Table 5) for the precision limits for IWTO-8 (projection microscope), IWTO-12 (Sirolan-Laserscan™), IWTO-28, (airflow) and IWTO-47 (OFDA). He concluded that differences between instruments, although in some cases statistically significant, are generally small, and often attributable to deficiencies in the airflow instrument. He reaffirmed the robustness of the projection microscope, despite its deficiencies, as a primary reference method for calibrating other instruments. He stated that the 76 references that he had cited only represented some 25% of all the published data at that time.

Baxter [227] and Sommerville and Teasdale [280] have analysed and summarised comparative studies between the various diameter-measuring instruments, noting the improved calibration techniques that have been introduced subsequent to the original studies. The differences that still remained were ascribed to the different definitions (measurement principles) of fibre fineness employed and implied by the different types of

Table 5.   Precision of the instruments used for determining the fineness of wool fibres [283]. Reprinted from P.J. Sommerville, Wool Technol. Sheep Breed. 48(3) (2002) pp. 198–232, with permission of IWTO.

Instrument	Precision (95% confidence level)	
	20 μm	35 μm
Projection microscope	±0.87μm	±1.07μm
Airflow*	±0.45μm	±0.80μm
OFDA**	±0.36μm	±0.67μm
Sirolan-Laserscan™**	±0.32μm	±0.70μm

Notes: *Sliver.
**Raw wool.

instruments [283,304,343,393], and this could be accommodated by specifying the measuring instrument and method in contracts etc. Various IWTO international round trials involving fibre diameter measurement have been held and reported [315,316].

As far as wool sliver is concerned, the different instruments, in terms of inter-laboratory variation, have been ranked as follows [169,383]:

- Projection microscope (highest variation)
- Airflow
- OFDA
- Sirolan-Laserscan™ (lowest/least variation)

Table 5 compares the precision of different instruments [283].

Different ways of improving the OFDA and the Sirolan-Laserscan™ measurements of the diameter of fibres extracted from slubbings, yarns and fabrics (IWTO-12, Appendix E), particularly with respect to the improved dispersion of fibre snippets, have been addressed [394–396] leading to an improved preparation and measurement technique [394] and within-laboratory variation. It is important that the calibration material is prepared in the same way as used for the fabric [394,397].

2.3.6. Interrelationships between mean fibre diameter, standard deviation, coefficient of variation and coarse edge

2.3.6.1. Introduction. It is important, in many instances, to establish the interrelationship between the different parameters used to characterise wool fibre diameter and its distribution, as well as typical (or average) values for such parameters. For example, it is known that fibre diameter variability (i.e. CV or SD) does play a small but significant role in spinning performance and yarn irregularity, and the question which frequently arises is whether the variability in fibre diameter, expressed as either SD or CV, is excessive for a particular mean fibre diameter raw wool or top. Similar considerations apply to the CE in terms of handle and prickliness. In order to answer such questions in a meaningful way, it is necessary to know the relationship between mean fibre diameter and SD, CV and CE, and to have appropriate 'average' or 'typical' values for different mean fibre diameters at the various stages in the wool processing pipeline.

2.3.6.2. Standard deviation versus mean fibre diameter. With respect to the relationship between SD and mean fibre diameter for wool, from greasy to top, the results of various studies have shown that SD is linearly related to mean fibre diameter for all practical purposes [179,189,275,344,398–412,416], although the scatter is fairly large, depending, amongst other factors, on the stage at which the measurements are made, e.g. fleece, bale, sale lot or top, as well as on breeding, classing, lot building and blending practices. The linear relationship between mean fibre diameter and SD [412] also extends to the changes in diameter along the length of a staple [344].

Greasy (raw) wool. Whiteley and co-workers [407] derived the following equation from measurements (FFDA) on almost 3000 auction lots of greasy wool:

$$SD = 0.263D - 0.325$$
$$(R^2 = 0.815). \tag{1}$$

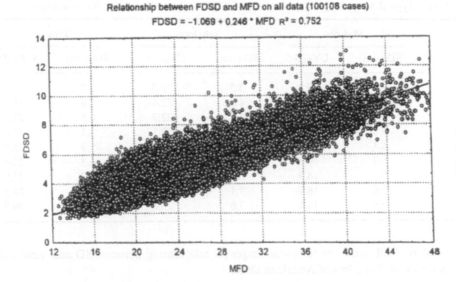

Figure 8. Relationship between FDSD (SD) and MFD (*D*) for combined mid-side samples [399]. Reprinted from B.P. Baxter and D.J. Cottle, Wool Technol. Sheep Breed. 46(2) (1998) pp. 154–171, with permission of IWTO.

Naylor et al. [398] derived the following empirical linear relationships from the measurements of 297 greasy wool auction lots (mean fibre diameter ranging from 19–25 μm):

$$SD = 0.23\,D - 0.29 \text{ (Sirolan-Laserscan)} \tag{2}$$

and

$$SD = 0.23\,D + 0.08 \text{ (FDA200)}. \tag{3}$$

Baxter and Cottle [399] obtained the following relationship using the results of some 100,108 mid-rib wool samples, obtained during the fleece test results on 1129 flocks from the 1996 season (samples originated from three broad geographic locations in Australia and New Zealand) and plotted in Figure 8:

$$SD = 0.246\,D - 1.069$$
$$(R^2 = 0.752). \tag{4}$$

The findings of Whiteley et al. [400,407] and Edmunds were confirmed, namely that adding a quadratic term to the regression equation only improved the fit by 1% and hence a linear relationship can be assumed without a significant loss in accuracy. Botha [118] found good agreement between mid-rib and fleece SD values for South African wools. Table 6 summarises the relationship between mean fibre diameter, CV_D and SD for wool measured at different stages. The values are the average values based upon the published work [118,344,398,399,401,403,404,407,410,413].

Table 6.   Typical SD and CV values for wool of different diameters and at different stages.

D (μm)	Mid-rib		Sale lots		Tops	
	SD (μm)	CV_D(%)	SD (μm)	CV_D (%)	SD (μm)	CV_D (%)
16	3.3	20.5	3.5	22.0	3.2	21.3
18	3.7	20.3	4.0	22.4	3.8	21.7
20	4.1	20.0	4.5	22.7	4.5	22.2
22	4.4	19.8	5.1	23.1	5.1	22.6
24	4.7	19.5	5.6	23.5	5.7	23.0
26	5.0	19.2	6.1	23.8	6.3	23.4
28	5.3	18.9	6.6	24.2	7.0	23.9
30	5.6	18.7	7.1	24.6	7.6	24.3
32	5.9	18.4	7.6	25.0	8.2	26.7

Lupton [410] found the following empirical relationship between SD and mean fibre diameter for 800 sale lots of American wool:

$$SD = 0.36D - 3.42$$
$$(R^2 = 0.95). \tag{5}$$

As would be expected, the SD of sale lots tends to be higher than that of the mid-side samples from which the sale lots were formed. From basic principles one would expect that for the same mean fibre diameter SD would progressively increase from staple to mid-side sample, to fleece, to bale, to farmer lots, to sale lot, to processing lot and to top. The actual magnitude of the increase will, however, depend upon many factors, such as genetic, environmental, classing (grower and bulk), clip preparation, sale lot and processing lot preparation and blending.

Tops. A number of studies have demonstrated a largely linear relationship between SD and mean fibre diameter for tops [275,399,401–404,408,409,411], and for different instruments, such as the projection microscope, FFDA, FDA200, Sirolan-Laserscan™ and OFDA. For example, in 1958 Ott [403] derived the following relationship between SD and mean fibre diameter for 5855 top samples, using the projection microscope, this still being the most extensive work done on tops:

$$SD = 0.346\ D - 2.42. \tag{6}$$

In the report by David [401], describing the relationship between the mean and SD of fibre diameter of wool tops, the following general equation was given:

$$SD = 0.345\ D - 2.403$$
$$(R^2 = 0.996). \tag{7}$$

Naylor [404] used the Sirolan-Laserscan™ method to measure the fibre diameter distribution of 100 top samples from four different mills, and obtained the following empirical relationship:

$$SD = 0.337\ D - 2.59$$
$$(R^2 = 0.98). \tag{8}$$

Table 7. Examples of linear regression equations drived for SD as a function of MFD (SD = constant + coeff × MFD) [399,411]. Reprinted from B.P. Baxter and D.J. Cottle, Wool Technol. Sheep Breed. 46(2) (1998) pp. 154–171 and B.P. Baxter and D.J. Cottle, *Fibre diameter distribution characteristics of midside (fleece) samples and their use in sheep breeding*, IWTO Report No. 12, Boston, USA, May 1997, with permission of IWTO.

Year	Author	Wool stale	Instalment	Number	Const. samples	MFD coeff.	CV (% at 20 μm)	CV (% at 30 μm)
1958	Oil	Tops	PM	5855	−2.42	0.346	22.5	26.4
1983	Hadwich	IH Tops	PM	49	−2.13	0.318	21.2	24.7
1983	Lunney	Tops	FFDA	41	−2.55	0.346	21.9	26.1
1984	David	Tops	General	Summary	−2.40	0.345	22.5	26.5
1991	Bow	Tops	FDA 200	222	−0.86	0.276	23.3	24.7
1993	Edmunds	IH tops	PM	n/s	−1.99	0.314	21.5	24.8
1994	Bow	Tops	Laserscan	156	+0.09	0.22	22.5	22.3
1996	Naylor	Tops	Laserscan	100	−2.59	0.337	20.8	25.1
1984	Whiteley	Sale lots	FFDA	2921	−0.33	0.263	24.7	25.2
1991	Bow	Sale lots	FDA 200	162	−0.35	0.244	22.7	23.3
1993	Edmunds	Sale lots	OFDA	228	−1.51	0.298	22.3	24.8
1995	Naylor	Sale lots	FDA 200	297	+0.08	0.23	23.4	23.3
1995	Naylor	Sale lots	Laserscan	297	−0.29	0.23	21.6	21.0
1982	Fleet	Fleece	FFDA	248	n/a	n/a	20.8	n/a
1994	Bow	Flc Mer	Laserscan	737	−1.19	0.25	19.1	21.0
1994	Bow	Flc Xbd	Laserscan	79	−0.23	0.20	18.9	19.2

It was also concluded that there was no real evidence in this respect that the nature of wool tops had changed over the last 38 years. Nevertheless, in practice it is possible that SD can differ for wools of the same diameter, but for different origins or periods, Baxter and Cottle [399,411] gave the following table (Table 7), which summarises the work done prior to 1986 on SD versus mean fibre diameter.

2.3.6.3. Coefficient of variation of diameter (CV_D) versus mean fibre diameter: Greasy (raw) wool. According to Section 2.3.6.2, SD increases more or less linearly with respect to mean fibre diameter. This has resulted in the CV_D becoming the preferred and convenient statistic when comparing the fibre diameter distributions of wools differing in mean fibre diameter, because CV_D is largely, if not completely, independent of mean fibre diameter and therefore provides a good basis for comparing the spread of diameter values for wools differing in mean fibre diameter.

As discussed in the previous section, Baxter and Cottle [399] obtained SD values for some 100,108 mid-rib wool samples. They also derived a common relationship between CV_D and mean fibre diameter for these mid-rib samples, which originated from three geographic locations in Australia and New Zealand. Statistically there were no significant differences between the three datasets, the average CV_D respectively being 19.2%, 19.5% and 19.1% for the Aust1, Aust2 and NZ1 datasets. For this reason, they only analysed and discussed the combined results and derived the following empirical relationship between CV_D and mean fibre diameter:

$$CV_D = 0.196\,D + 15.2$$

$$(R^2 = 0.078). \tag{9}$$

It is clear from the above relationship that CV_D is essentially independent of mean fibre diameter, particularly when the latter is below 26 μm, implying that CV_D can be assumed to be largely independent of mean fibre diameter for Merino-type wools. It was shown that for the fine end of the spectrum, which dominated their dataset, the average CV_D was typically 18–19%. Nevertheless, the 95% confidence limits at 20 μm covered a range from 13 – 25%, suggesting that there was considerable scope to reduce fibre diameter variability in many flocks, taking into consideration genetic, sickness and nutritional factors, which could influence CV_D.

On the basis of the results obtained on the same 297 auction lots mentioned in Section 2.3.6.2 and measured with the Sirolan-Laserscan™ method, Naylor and co-workers [398, 413] analysed the commonly occurring range of CV_D (18 to 28%) as a function of mean fibre diameter, 95% of the CV_D values falling within the range of 19 to 26%. Over this limited diameter range (19 to 25 μm), CV_D was independent of mean fibre diameter, having an average value of 22.2%. According to the work done on sale lots covering the period of 1991 to 1995 by various people [399] and taking into consideration the difference of about 1% in CV_D between the FFDA or FDA200 instruments and the Sirolan-Laserscan™, the average CV_D for the sale lots was also 22% for 20 μm wools. This implies that no real differences in fibre diameter variation occurred over the four years in question.

On the basis of the various results [344,398,399,405,413–415] it can therefore be concluded that CV_D is reasonably independent of mean fibre diameter for mid-rib and sale lots finer than about 25 μm, i.e. Merino-type apparel wools, with CV_D possibly increasing slightly with an increase in mean fibre diameter [435] by approximately 1% (absolute) for a 1 μm increase in mean fibre diameter for sale lots [10,398,417], and approximately equalling mean fibre diameter + 1%. For sale lots, CV_D is on average about 22% for a 20/21 μm wool, while it is about 18 to 19% for mid-rib samples [399,411].

Wool tops. The diameter distribution of 120 commercial tops from six different Australian top makers was measured, and no significant differences were found between the different top makers in terms of the average width (CV_D) of the fibre diameter distribution for a given mean fibre diameter [345,413]. It was found that for a given mean diameter, the range of CV_D values for the tops is, on average, about 1.5% narrower, in absolute terms, than the range for the sale lots. The following linear regression equation was obtained [413] for sale lots as well as for tops:

$$CV_D = 0.5\,D + 10.7$$

$$(R^2 = 0.60,\ \ SE = 1). \tag{10}$$

As with the sale lot data and taking into account the instrument differences [398], these results agreed well with the earlier results of Bow and David [402]. This relationship leads to the simple rule of thumb that a 21 μm wool top would be expected to have a CV_D value of 21 ± 0.5% and that CV_D increases on average by about 0.5 percentage points per micrometer increase in mean fibre diameter. Baxter and Cottle [411] stated that the average predicted CV_D for 20 μm wool would be 19%, 22% and 22% for fleeces, sale lots and tops, respectively. Botha [118] found CV_D to be independent of mean fibre diameter for tops produced in South Africa.

General. It can be concluded that there is essentially a linear relationship between SD and mean fibre diameter, whereas CV_D only increases very slightly with increasing fibre diameter, between approximately 0.5 percentage point and 1 percentage point for each 1 μm increase in mean fibre diameter for sale lots and tops. CV_D is virtually constant, between 18% and 19% for mid-side Merino fleece samples, whereas it is, on average, approximately 21% for 21 μm sale lots and tops.

2.3.6.4. Coarse edge versus mean fibre diameter, standard deviation and coefficient of variation. Introduction. Earlier in Section 2.1.2, various types of distributions were discussed along with different parameters that can be used to define them. It was also noted that the fibre diameter distribution curve could be represented by either a lognormal or Gaussian (i.e. normal) distribution. The coarse side (end) of the distribution, notably the percentage of fibres coarser than 30 μm (referred to as the CE or prickle factor (PF) [418–430], with a new definition as used today, developed by Knowles [431] and Marler [432] and named comfort factor (CF)) is of particular interest because of its major effect on prickliness or scratchiness.

Although, as would be expected, there is a broad relationship between CE and mean fibre diameter, the relationship is not sufficiently precise for prediction purposes, it being necessary to include a measure of diameter dispersion (distribution) in addition to mean fibre diameter. Nevertheless, for purposes of reference and comparison, Botha [118] and Botha and Hunter [433] created a new term, namely the Coarse Edge Ratio (CE$_R$), which is the ratio of the actual (measured) CE to the 'average' or 'typical' CE value for wool of the same diameter. If CE$_R$ is 1, it indicates that the particular wool has the CE expected for an 'average' or 'typical' wool of that diameter, and if CE$_R$ is above 1, it indicates that the wool has an 'above average" CE level etc.

Greasy (raw) wool. Whiteley and Thompson [434] and Lunney [189] have reported on the coarse edge statistics of sale lots of Australian greasy wool, concluding that mean fibre diameter and SD provided an adequate indication of the proportion of coarse fibres in greasy sale lots. Phillips [345], using the results of the study done on 1397 Australian sale lots by Whiteley et al. [400], showed that by applying the Gaussian distribution theory, the relationship between CE and SD, and CV$_D$ and mean fibre diameter, can be used to calculate the CE. Baxter and Cottle [399] measured the CE and found that there was no simple way of predicting CE in practice without taking into account both mean fibre diameter and either SD or CV$_D$. When relating the variation in CE to mean fibre diameter on its own (covering a range of 12.6 to 52 μm) using the power law relationship as proposed by Naylor [404], a regression of the following form was obtained [399]:

$$CE = 3.48 \times 10 - 14(D)10.45$$
$$(R^2 = 0.792, \ SE = 2.0, \ N = 93839). \tag{11}$$

Applying the standard Gaussian score, Z, which incorporates mean fibre diameter, SD and CV$_D$, the following more robust and precise prediction of CE was derived for the combined dataset [399]:

$$CE = 17.30 - 4.21$$
$$(R^2 = 0.943, \ SE = 1.0, \ N = 93839). \tag{12}$$

Whilst Equation (12) was a more cumbersome formula to apply, the precision of prediction of CE was much better and was relatively uniform throughout the critical CE range (i.e. from 2 to 6%). Baxter and Cottle [399] also showed that flocks having a 5% CE level could have CV$_D$ values ranging from 15 to 25% within the mean fibre diameter range of 21 to 23 μm.

The selection of sale lots based on SD or CV$_D$ alone is insufficient for selecting wools with a CE below 5%. However, simple selection rules can link SD and mean fibre diameter

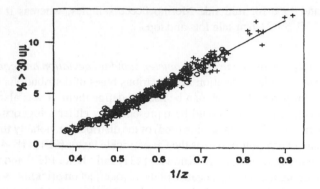

Figure 9. A plot of CE vs. $\frac{1}{z}$. The solid line represents the line of best fit and the symbols + and O correspond to the CV_D values greater and smaller than 22.2% [398], respectively. Reprinted from G.R.S. Naylor, D.G. Phillips and C.J. Veitch, Wool Technol. Sheep Breed. 43(1) (1995) pp. 69–82, with permission of IWTO.

fairly accurately to define wools with potentially acceptable coarse fibre levels. According to the derived relationship, a 21 μm wool with a CV_D of 25.9% is equivalent in terms of CE to a 22 μm wool with a CV_D of 22%, or a 1 μm shift is equivalent to a 3.2% change in CV_D at 24 μm, increasing to a 4.8% change in CV_D at 19 μm. Phillips [345] showed that in terms of CE levels one-micron shift in diameter is equivalent to a change in CV_D of 3 to 5% and that careful selection of low prickle wools from sale lots only needs to concentrate on wools within the 21 to 22 μm groups. For typical Australian fleece wools, 3 to 8% shift in CV_D is equivalent to a 1 μm change in mean fibre diameter in terms of their effect on CE. The lower values occur in the region of low mean fibre diameter and low CV_D and increase with an increase in mean fibre diameter and CV_D.

Naylor et al. [398] confirmed the results of Phillips [345] on sale lots measured by the Sirolan-Laserscan™ method (mean fibre diameter ranging from 20 to 24 μm and CV_D from 18 to 28%). As a rule of thumb, proportion of fibres greater than 30 μm was determined from the combination of the standard Gaussian transformation for CE and the line of best fit from the sale lots (see Equations (13) and (14) respectively):

$$Z = \frac{30 - D}{SD} = \frac{30 - D}{D.CV_D} \times 100, \tag{13}$$

$$CE = 21.9\frac{1}{Z} - 7.51$$
$$(R^2 = 0.963). \tag{14}$$

Figure 9 [398] represents the above equation of CE as a function of $\frac{1}{z}$ for the sale lots, which is conveniently linear over the region of interest. Thus, wools with the same Z values have very similar CE values. As can be expected, for a given mean fibre diameter, samples with a larger than average CV_D also have a larger than average CE. The following best fit equation between CE and mean fibre diameter was obtained:

$$CE = 2.09\,D - 40$$
$$(R^2 = 0.735). \tag{15}$$

It can be seen that the R^2 value is much greater for Equation (14) than for Equation (15).

Combining Equations (13) and (14) gives the following relationship:

$$CE = \frac{0.219.D.CV_D + 7.5.D - 225}{30 - D}. \tag{16}$$

Barker and Fookes [435] found a close association between CE and mean fibre diameter for New Zealand's relatively fine wool (15 to 24 μm) sale lots.

Botha [118] found a close association between CE and mean fibre diameter for South African fleece wools, with a more hairy birth-coat being associated with higher CV_D and CE levels, particularly for 2-month-old lambs.

Tops. As already noted, the magnitude of the coarse fibre tail in the diameter distribution is dependent on the intrinsic shape of the distribution as well as the mean and coefficient of variation. The standard Gaussian distribution, with the associated mean fibre diameter and SD, is for most purposes sufficient to describe the diameter distribution of commercial tops.

Applying the theory of the standard Gaussian score to the some 100 commercial tops, Naylor and co-workers [344,404,436] showed that within the diameter range 20 μm $< D <$ 24 μm, a linear relationship between CE and $\frac{1}{z}$ was obtained, with the best fit line ($R^2 =$ 0.98, MSE = 0.07) being

$$CE = 19.97\frac{1}{Z} - 6.79. \tag{17}$$

Naylor [404] and Naylor and Phillips [436] showed that by applying log transformation to the data, a straight line could be fitted to the data for the full diameter range ($R^2 = 0.96$, MSE = 0.05), with the following relationship:

$$\log_e CE = -27.8 + 9.5 \log_e(D), \quad \text{i.e.} \tag{18}$$

$$CE = (8.4 \times 10^{-13})D^{9.5} \tag{19}$$

Similar results were obtained by Botha [118] for South African Merino-type wools with the following relationship:

$$CE = (1.0 \times 10^{-9})D^{7.2}. \tag{20}$$

Naylor [404] and Naylor and Phillips [436] found that there was a linear relationship ($R^2 = 0.98$, MSE = 0.09) between the mean diameter of the fibre ends and the usual mean diameter, with the fibre ends being consistently finer (on average about 1 μm finer) than the usual mean fibre diameter for both raw wool and tops [437]. Naylor [438] discussed ways in which 'fine fibre ends' tops could be produced, while Mahar et al. [439] presented a preliminary report on the prediction of the fibre end characteristics in the top from the raw wool characteristics.

2.3.6.5. Changes in mean fibre diameter, fibre diameter distribution and coarse edge during processing. Mean fibre diameter and fibre diameter distribution change little from the raw wool to the fabric [29,440–445], although finer fibres break more easily during processing.

Combing removes shorter fibres, which are generally finer [34,446–453], with the result that the top tends to be marginally coarser (about 0.2 to 0.6 μm) [34,112,179,452,454–458] than the raw wool, depending upon the mean diameter [34,458,459]. Bell [112] reported that in order to predict the top mean fibre diameter from that of the raw wool, an allowance on a sliding scale from +0.2 μm at 19 μm to +0.6 μm at 27 μm is made, the top always being coarser.

Analysis was undertaken by Turpie and Gee [454] on some 250 lots of South African wools ranging from 18 to 30 μm in mean fibre diameter, with a VM base averaging about 1% and generally of good/average top-making style which were processed on pilot scale machinery. They examined the correlation between the mean fibre diameter of the top and that of the raw wool and found the best fit (explaining some 96% of variability) to be given by the following quadratic equation [454]:

$$D_T = 4.9 + 0.62 D_W + 0.007 D_W^2, \qquad (21)$$

where D_T is the mean fibre diameter of the top and D_W is the mean fibre diameter of the raw wool.

From the above equation it follows that the difference in diameter (mean diameter of top minus that of greasy wool) decreased from +0.3 μm at 18 μm to a minimum of −0.3 μm at 27 μm and increased again to +0.3 μm at 36 μm.

According to Naylor [404] the various wastes generated during processing prior to the top have higher CE values than the tops, whereas the noil removed during combing generally has a lower CE than the tops, confirming that card waste tends to contain more coarse fibres and the noil fewer coarse fibres than the top due to the removal of relatively fine fibres as noil during combing [447].

A report by Bell [460] to the Raw Wool Service Department of the International Wool Secretariat (IWS) in 1975 gave information on diameter distribution of six blends of superfine merino greasy wool lots (average fineness 18.5 μm) and on the diameter distribution of the resultant tops. He showed that both mean fibre diameter and SD were consistently greater for the tops than for the corresponding greasy wools, the overall difference in mean fibre diameter being not statistically significant, while that for SD it was highly significant. Pohle et al. [452] investigated the relationships between the fibre diameter and diameter distribution of greasy wool and those of the top and found that the average fibre diameter of the top was greater than that of the noil, card sliver and greasy wool. Their results on noil showed that, although the greatest percentage of fibres removed during combing were the finer ones, there were also fibres in the noil covering the entire range of fibre distribution found in the top. For the lots tested, there was a general tendency for the SD and CV_D of the top to be slightly higher than those for the corresponding card sliver and greasy wool. Research work done by Kurdo et al. [447] largely confirmed the above results, showing that the noil was 0.3 μm finer than the top, card sliver and greasy wool, while the mean fibre diameter and SD of the card waste were higher than those of the greasy wool, top and card sliver. Butler et al. [461] concluded that a reasonable quantity of fibres get lost as waste which could alter the diameter distribution of the resultant top. Baxter and Cottle [411] found the CV_D of sale lots and tops to be the same and slightly higher (e.g. 22% vs. 19% for 20 μm wool) than that of the fleeces, in line with the Quinnell et al. analysis [223]. Other workers [462] reported that commercial tops had higher variability in diameter than found in the case for Australian flocks. In other studies [89,463], the coefficients of variation in diameter of tops were found to be slightly lower than those of the sale lots that were processed into the tops, ascribed to the removal of relatively fine fibres as noil during combing.

Botha [118] found the CE levels in sale lots and commercial tops to be similar and generally higher than those in mid-rib samples of the same mean fibre diameter, indicating the potentially adverse effect of commercial blending practices on CE levels. Nevertheless, while processing fleeces into tops on a pilot plant, the CE levels were found to be similar at the different processing stages, while CE and CV_D tended to be higher for the sweepings, strippings, noil and shoddy than for the fleeces, card slivers and tops; both for the skirted and unskirted fleeces processed into tops [118].

David and Andrews [453] concluded that when a top is produced from a batch of raw wool, the most likely values of the parameters of fibre diameter distribution of the top are the corresponding parameters of the raw wool itself. Data from the trials evaluating additional measurement (TEAM) [464] confirmed this trend and showed that differences between greasy core and top diameter were statistically nonsignificant in contrast to various other studies which indicate that tops tend to be slightly finer than the raw wool from which they were produced. It was also postulated [453,465] that the differences in greasy wool and top diameter values observed in the past could have been due to differences in the airflow calibration required for tops and greasy wool, respectively, i.e. due to differences in measurement rather than due to actual differences in diameter. Differences in processing conditions may, however, have played a role.

Botha [118] found a similar relationship between CE and mean fibre diameter for tops produced on a pilot plant and those produced commercially (Figures 10 and 11). He found very little differences between different breeds of sheep and different topmakers (Figures 10 and 11).

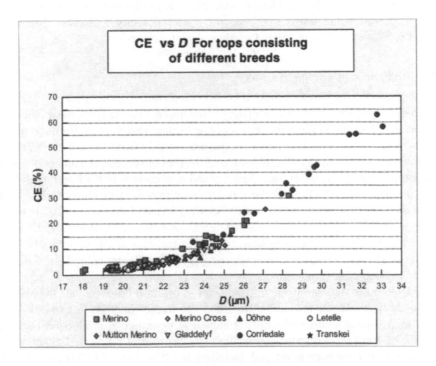

Figure 10. CE vs. *D* for tops representing different sheep breeds (CSIR pilot plant lots).

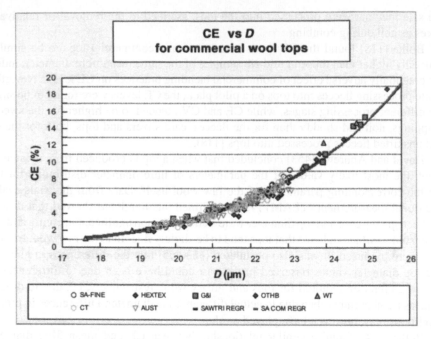

Figure 11. CE vs. *D* for commercial tops.

Summary. Taking an overall view of the results of various studies it can be concluded that there is only a small difference between the mean fibre diameter of a top and that of the greasy wool from which it was produced, with the top possibly marginally coarser due to the preferential removal of relatively fine fibres during processing.

2.3.7. *On-farm and staple profile measurements*

Although fleece testing has been carried out for a very long time, this has generally involved testing mid-side samples in a central laboratory [466] and not on the farm, the relationship between fleece test results and core test results having been reported [467]. The quest for on-farm measurement of wool quality-related properties, notably diameter and its variability, has been pursued over many decades with varying degrees of success. A number of papers have discussed the purpose, advantages, benefits and applications of flock testing and on-farm measurement [468–482], particularly in terms of improved classing, breeding and other farm management practices, as Lamb [85,483] stated, summarising the potential benefits of on-farm measurement of individual fleeces. Atkins and Semple [484] discussed the on-farm testing of fibre diameter for objective clip preparation and its economic benefits, while Andrews et al. [475] reported on the classing of Merino fleeces using fibre diameters predicted from hogget measurements.

Earlier attempts to measure wool fibre diameter of on-farm fibres include the Mikron-meter [244–249] developed in South Africa and the Sonic Fineness Tester [235–238] developed in Australia, neither of which remained in commercial production for long. Recently, the on-farm measurement of fibre diameter has received increased attention and various instruments have been developed, including the WoolView 20/20 (Figure 12), Dr Kurup's Wool Micron Tester (Figure 13), OFDA2000 (Figure 14), Sirolan-Fleecescan™

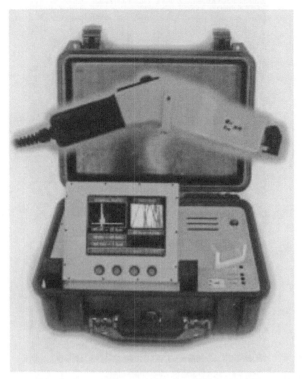

Figure 12. WoolView 20/20 (www.woolview.com), reprinted with permission of Mark Brims.

(Figure 15) and the FibreLux (Figure 16), none of which are intended or used for certification or commercial trading purposes as yet. The use of such testing instruments for on-farm measurement allows growers to benefit not only from the instantaneous availability of data important for farm management and animal selection purposes but also from being able to prepare accurately classed and uniform lots of objectively measured (diameter) wool for sale. Of the above-mentioned instruments, the two instruments that have featured most by far in the technical and scientific literature are the OFDA2000 and the Sirolan Fleecescan™, and are discussed in more detail below.

Brims et al. [485] introduced the OFDA2000 (presented to the IWTO in 1999 and launched in 2000) for the rapid on-farm measurement of the diameter profile of greasy wool staples. Baxter [486] described the OFDA2000, including its accuracy and precision compared to conventional instruments and its ability to provide staple length and length–diameter profile data which could be used in various ways, even for processing prediction models. The OFDA2000 measures the fibre diameter characteristics along the length of greasy micro-staples prepared from a staple drawn from a consistent site on each fleece, usually the mid-rib or pin-bone [334,466,485–493] (www.ofda.com). Generally, one micro-staple is measured, Brown et al. [494] finding that one staple was not sufficient to reliably determine differences between sheep, but was sufficient to obtain a reliable average for a group of animals. The instrument corrects for variations in ambient temperature and humidity and a flock-specific correction is applied for grease. The grease correction factor for the OFDA is about 1.4 μm on average, but varies between flocks and is diameter-dependent [490]. Typically, a 0.8 μm thick layer of grease and suint coats a 20 μm wool

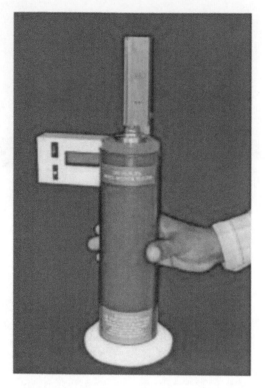

Figure 13. Dr Kurup's wool micron tester (from manual), reprinted with permission of Dr. Vinod Kurup.

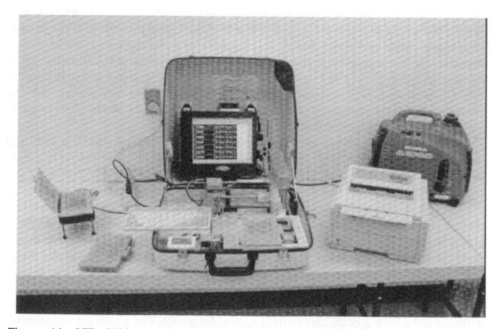

Figure 14. OFDA2000 (www.awta.com.au/upload/Presentations/On_Farm. . . /On_farm_Testing. ppt), reprinted with permission of AWTA Ltd.

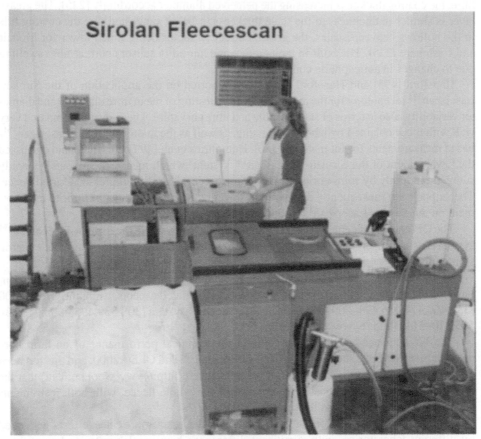

Figure 15. Sirolan-Fleecescan™ (www.csiro.au/solutions/SirolanLaserscan.html), reprinted with permission of CSIRO.

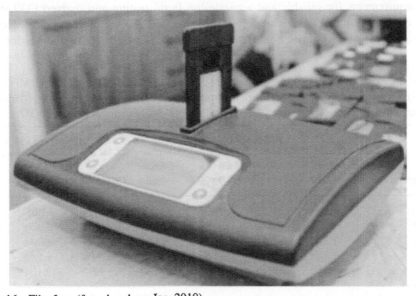

Figure 16. FibreLux (from brochure Jan. 2010).

fibre, the coating thickness increasing the perceived diameter accordingly [225]. The grease layer is similar in thickness to the scale thickness, it being suggested that the coating fills in the hollows between scales, the grease correction factor being much lower for Mohair and Cashmere [225]. The built-in temperature and humidity sensors correct fibre swelling due to changes in atmospheric conditions.

Hansford [495] and Hansford et al. [496] reported on the application of the Sirolan-Laserscan™ technology for the real-time-on-farm testing of mean fibre diameter and diameter variability, also in terms of staple profile and fibre curvature. These results compared very well with those obtained by laboratory testing, as well as the in-store Sirolan-Laserscan™-based measurements for lot matching [80]. Humphries et al. [497] presented and discussed the development of the Sirolan-Fleecescan™ system, which measures snippets generally sourced randomly by mini-coring (2 mm diameter core) a roll-compressed skirted fleece. The snippets are subsequently degreased in solvent (tetrachloroethylene) prior to measurement on the Sirolan-Laserscan™ [466] system.

The OFDA2000 is generally used for pre-shearing tests, while the Sirolan-Fleecescan™ system is usually operated in the shed, during shearing, or possibly post-shearing at a central store [466].

Many studies have been undertaken on the applications, precision and accuracy of on-farm testing of diameter and diameter profile and the related aspects, using the OFDA2000 [339,466,474,486,498–509] and Sirolan-Fleecescan™ system [308,339,495,498,501,503–505,510–514] (http://www.awta.com.au).

The AWI financed an extensive study of the technical performance of on-farm fibre measurement systems, such as the Sirolan-Fleecescan™ and OFDA2000, and this has been reported on http://www.awta.com.au. It was concluded (http://www.wool.com.au) that the mid-side sample was the most representative of the whole fleece, with both instruments delivering a similar precision for such a sample.

Morgan [510] reported on factors influencing the precision of fleece tests generated by different laboratories, with significant differences between laboratories having been reported.

Marler and co-workers [474,498,501,503,515] reviewed and reported on the reasons and requirements for reliable and accurate 'on-farm' testing and derived [339,498,503] 95% confidence limits (see Table 8 [503]) for the OFDA2000 and Sirolan-Fleecescan™ measurements of the diameter characteristics of fleeces. Marler [516] and Marler and co-

Table 8. 95% whole fleece confidence limits for MFD, CVD, SDD, CFR and MFC for the properties measured using various instruments/methods [503]. Reprinted from J.W. Marler and P. Baxter, Wool Technol. Sheep Breed. 52(1) (2004) pp. 43–96, with permission of IWTO.

Parameter	Fleece Whole flc avg	Fleece Lab LSN	Pinbone samples Lab OFDA100	Pinbone samples OFDA2000	Fleece Fleece-scan	Mid-side samples OFDA2000	Mid-side samples Lab OFDA100	Mid-side samples Lab LSN
MFD	19.5	±1.19c	±1.25c	±1.39d	±1.17bc	±1.24c	±1.05ab	±1.04a
SDD	4.1	±0.7a	±0.5a	±0.5a	±0.7a	±0.5a	±0.5a	±0.6a
CV_D	20.8	±3.3c	±2.4a	±2.2a	±3.4c	±2.2a	±2.4a	±2.9b
CFR	98.1	±2.3a	±2.6a	±2.6a	±1.8a	±1.6a	±1.5a	±1.6a
MFC	93	±15b	±12ab	±13b	±13ab	±12ab	±10a	±13ab

Note: For a given parameter, values with the same letter are considered to be statistically equivalent ($p > 0.05$).

workers [339,474] have noted that it is the precision of the entire fleece and not that of the single sampling site (e.g. pin-bone or mid-side) that is important in practice; the values between 0.6 μm and 1.6 μm given for the confidence limits for fleece testing [79,489,510] tend to underestimate the precision limits relevant for making 'on-farm' decisions [515]. Marler et al. [498] estimated the 95% confidence limits for mean fibre diameter to be ±1.41 μm for OFDA2000, ±1.02 μm for Sirolan-Fleecescan™ and ±1.19 μm for the Lab-Laserscan systems. Marler and Baxter [503] concluded that the OFDA2000 diameter values for the mid-side samples and the Sirolan-Fleecescan™ results obtained on the whole fleeces gave equivalent accuracy compared with the fleece average, the pinbone samples giving values on average 0.6 μm higher than the fleeces. Atkins [517] found the precision of on-farm measurement of diameter to be related to the mean fibre diameter. Baxter and Marler [504] concluded that the two on-farm sampling and measuring technologies produced very similar outcomes, with only slightly lesser precision than laboratory fleece testing on mid-side sampling in accordance with AS/NZ 4492. Hansford et al. [502], however, recommended that direct comparisons of results from OFDA2000 and Sirolan-Fleecescan™ systems should not be made.

The measurement of fibre diameter profile (i.e. along fibres and staples) using various techniques, notably the OFDA2000, and the implications thereof on staple strength and its improvement, farm management, processing etc. have been reported by various workers [223,287,507,518–533]. It has been suggested [533] that staple profile could perhaps provide a substitute for staple strength because they were highly correlated, although certain workers have suggested that CV_D was preferable to other measures for selection and breeding purposes.

The effect of various growth factors on staple profile has been studied [534,535], the magnitude of the effect of such growth (environmental) factors being determined genetically. Smith et al. applied the fibre diameter profiles for improving fine wool quality.

Although the on-farm measurement techniques were developed for farm management purposes, notably breeding and classing, there have also been attempts to relate such measurements, particularly those of the OFDA2000, to processing performance, notably for the prediction of Hauteur. Various workers proposed the use of the staple fibre diameter profile and staple length for processing prediction purposes, such as Hauteur [536], or for improving processing prediction formulae [486,521,537–541]. Peterson and Oldham showed, for example, that top properties could be predicted from the OFDA2000 fibre diameter properties. Hansford [537] suggested that the diameter profile of a staple is useful for predicting the position of break and possibly the processing performance, noting that fleeces with a relatively simple profile predicted better than those with a more complicated profile. The use of the OFDA staple profile measurements for predicting Hauteur, without the need for other staple measurements, has been mooted [536]. Nevertheless, Marler and Couchman [474] discussed and cautioned against such claims. Some of the concerns that were raised by Marler and Couchman [474] were also addressed by Peterson [509] and Cottle [508].

Many studies [542] have shown a strong relationship between fibre diameter variability, notably the along-fibre component, and fibre and staple strength, more particularly when no correction is made in strength in terms of the minimum fibre (or staple) cross-section or that at the point of break. Yamin et al. [482] concluded that the coefficient of variation of fibre diameter derived from the mid-side sample was as good as, if not better than, the along-staple values in terms of their usefulness for selection for improved staple strength. Hilton and Gardiner [542] discussed the application of the measuring techniques using the Fourier transforming property of a lens for the 'on-farm' measurement of fibre diameter.

Brown et al. [543] reported on the use of reduced profiling techniques to estimate the fibre diameter profile.

It should also be noted that other techniques, but not on-farm, have been used to measure along-staple cross-section and fibre diameter profiles. For example, the SAWTRI Automatic Staple Length and Strength Tester developed by Turpie et al. [544] measured the staple profile by means of a fluted roller (tongue) and groove device from which the staple length and position of break were derived. Various staple profiles were demonstrated, which after standardisation showed close agreement with the top Hauteur diagrams for sound wools. Grignet et al. [545] reported on the use of a modified sensor to evaluate the length and variation in linear density (cross-section) of raw wool. In 1985 Hansford et al. [546] reported a method to measure the diameter profile of wool staples using a staple cutter, which can cut wool staples into consecutive 2 mm snippets and then using the FFDA to measure the staple diameter profile. Schlink et al. [547] showed that the staple tex (i.e. linear density or cross-section) profile along the staple (either greasy or clean) was a suitable alternative to fibre diameter profile, the former predicting the latter, the precision being better for clean staples than for greasy staples.

2.4. Single fibre and staple length

2.4.1. Introduction

Fibre length is probably second in importance after diameter in terms of determining 'sound' apparel-type wool quality and performance, the fibre length characteristics of wool, notably the mean fibre length, playing a very important role in determining the processing route and performance, including fibre breakage, processing waste, spinning performance and limits, as well as the yarn tensile, evenness and hairiness characteristics. Dircks [548] has discussed the significance of fibre length in processing. It is important to note that fibre breakage during processing, particularly in the case of fine, long and tender wools, can greatly change the fibre length characteristics from the greasy to the top stage.

Greasy wool staple length and fibre length are largely determined by the period of wool growth (i.e. period between shearing) and by the breed of sheep (i.e. genetics), with environmental (e.g. nutrition) conditions playing a secondary role in practice. Nolan [11] reported that the stress of lactation reduced staple length by about 6%, common sheep diseases also reducing the staple length.

The overall average of staple length of Australian wool does not vary significantly from year to year [17], it being around 90 mm per 12-month growth, although it does vary from region to region (state to state), with finer bloodlines tending to have shorter staple lengths, for example 85 mm for fine wool hogget ewes, against 94 mm for medium and strong wool bloodlines. In addition to the above, the staple length, and also single fibre length, varies according to the position on the sheep. Within a staple and fleece there tends to be a positive relationship between fibre length and fibre diameter [549,550]. There also tends to be a positive relationship between fibre/staple length and fibre diameter [23], and between sheep, particularly between different breeds of sheep. The ratio of fibre length growth (L) within a certain growth period to mean fibre diameter, i.e. (L/D), varies widely between sheep, from about 10 to 20, but remains fairly constant with change in wool growth [551].

Various methods for determining wool fibre length have been developed over the years; certain of these being captured in various IWTO test methods and draft test methods, some of which are no longer in use. Fibre length of wool can be measured and described in various ways [552–555] and at various stages, from the greasy stage to the sliver or top stage. At the greasy stage, staple length, rather than fibre length, is measured during the objective measurement (pre-sale additional measurement) of wool for trading purposes, it being a lot easier and quicker to representatively sample and measure staples than single fibres. It is extremely difficult to remove, without fibre breakage, and measure accurately and cost-effectively single fibres from greasy staples. This is somewhat easier, but still difficult to do, when the wool has been scoured (using, for example, the New Zealand LAC method), the entanglement of fibres and therefore fibre breakage during fibre preparation for testing, presenting a problem. Furthermore, it is also not always easy to obtain a representative sample in this way.

The following are current IWTO test methods or draft test methods relating to the measurement of fibre length characteristics in wool sliver and top and staple length of raw wool:

- Method of determining 'Barbé' and 'Hauteur' for wool fibres using a comb sorter – Draft TM-1–02.
- Method of determining wool fibre length distribution of fibres from yarns or fabrics using a single fibre length measuring machine – Draft TM-5–97.
- Method of test for wool fibre length using a WIRA fibre diagram machine – Draft TM-16–02.
- Determination of fibre length distribution, mean fibre diameter and fibre diameter distribution of wool top and slivers by the OFDA – 4000 – Draft TM 62–01.
- Determination of fibre length and distribution parameters, almeter – IWTO-17–04.
- Determination of staple length and staple strength – IWTO-30–98 (amendment approved and test method issued in July 2007).

Single fibre length was traditionally measured manually on instruments, such as the WIRA single fibre length tester (IWTO-Draft TM-5 – 97), but this is done less frequently today because of the labour-intensive, expensive and time-consuming nature of the test.

WRONZ has developed a 'length after carding' (LAC) method for measuring the fibre length of scoured or carbonised wool, which involves a laboratory carding system that emulates a semi-worsted line [131]; a representative scoured wool sample (1 kg) is conditioned, lubricated, carded and gilled under standard conditions and then measured on an Almeter for mean fibre length (Barbé) and other length characteristics. This test became a full New Zealand Standard (NZS 8719) in 1992, an extension being the Card Waste Test [556], which assesses the propensity of wool to produce droppings and fly [557]. Wear and Baxter [558] have presented LAC results for some 3000 consignments of New Zealand wool, and found that CV Hauteur did not perform well as an indicator of short fibre content and that

$$\text{Hauteur (mm)} = 0.7\ \text{Barbé} + 2.$$

Baxter and Wear [559] concluded that the greasy LAC model provided the potential for more precise prediction of mill performance for New Zealand fine to medium wools, than either the TEAM or CSIRO TopSpec models, but this needed to be verified by mill or pilot plant processing trials.

2.4.2. Greasy staple length

According to IWTO 30, staple length is defined as 'the length of a staple projected along its axis obtained by measuring the staple without stretching or disturbing the crimp of the

Figure 17. ATLAS, Automatic staple length and strength tester (http://www.awta.com.au/en/Home/Publications/Newsletters/), reprinted with permission of AWTA Ltd.

fibres'. There is a fairly good relationship between staple length (particularly unstretched) and fibre length in the top, although the actual relationship depends upon fibre entanglement during scouring, fibre fineness, crimp, length, strength, fibre breakage during processing etc. The ratio of mean staple length to mean top fibre length (Hauteur) therefore varies accordingly, but, on average, is 1.2:1 [560]. The use of decrimped staple length to differentiate typical wool types as an aid to processing prediction has been investigated by Fish et al. [561].

Because of the difficulty in measuring the single fibre length characteristics of greasy wool, various instruments have been developed to automatically measure the staple length of greasy wool. Such instruments usually measure staple strength and staple length at the same time, and include the ATLAS (Automatic Testing of Length and Strength) (Figure 17) developed by the CSIRO in Australia and the SAWTRI Automatic Staple Length/Strength Tester (Figure 18) developed at the CSIR in South Africa [562–564], only the former being commercially available at present. Manual and semi-automatic staple length/strength testers, such as the Agritest Staple Breaker (Figure 19), have also been developed. All the above instruments are described and discussed in more detail in Section 2.4.6. Staple length and strength measurements were first introduced commercially for Australian combing wools in January 1985 [173] and soon afterwards in South Africa.

For objective measurement and trading purposes, nowadays greasy staple length is almost exclusively measured, together with staple strength, on the ATLAS automatic staple length/strength testers, according to IWTO-30–98 (amendment approved and test method issued in July 2007). Essentially this involves taking grab samples from the bales of wool (IWTO-38–99), drawing representative staples from such grab samples (IWTO-7–00) and then passing at least 60 staples (unstretched) lengthwise through a light beam, usually by means of a conveyor belt; the length of time for which the light beam is interrupted provides

Figure 18. SAWTRI automatic staple length/strength tester (SAWTRI brochure).

the measure of the staple length. After this, each staple is tested for strength and position of break, with the relative weight of the broken segments used to determine the 'position of break' (POB), after appropriate corrections for differences between the tip and root portion of the staple (see also Section 2.4.6).

2.4.3. Relationship between staple length and fibre length within the staple

The relationship between the mean length of fibres in a wool staple and the staple length varies considerably, their ratios (i.e. mean fibre length to staple length) reportedly varying from about 0.85 to 2.1 [35–37,565–572], more typically from 1.2 to 1.4 [573] or 1.0 to 1.3 [574], with an average of about 1.17 [575]. On the basis of the latter, the top Hauteur should be 1.17 times the staple length if no fibre breakage occurs during processing [575]. Techniques relying on the selection of the limited number of individual fibres from staples may not give an accurate representation of the staple fibre populations [576,577]. Delfosse et al. [574] made the assumption that the fibre length distribution within a staple is roughly log-normal. Whan [36] concluded that the use of staple length, as a predictor of the average fibre length of greasy wool, is not always reliable, whereas other workers [36,568,578–581] showed that within a flock staple length provided a reasonable measure of the mean fibre length within the staple. For example, Lang and Chaudhri [568] showed that the stretched staple length bears an approximately constant ratio to both the mean fibre length within

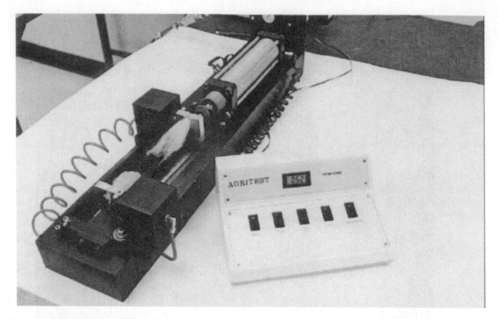

Figure 19. Agritest staple breaker (http://sanangelo.tamu.edu/wmrl/Lupton%20Presentations/Alpaca%20Information%20Day%20VSU%201-17-09.pdf), reprinted with permission of IWTO.

the staple and the crimped length of the staple. Bow [581] used results from three different sources to confirm the linear relationship between the mean fibre length in a staple and the mean staple length of greasy wool (Figure 20) [581].

Turpie and Cizek [563] found that the mean fibre length results obtained from single fibre measurements and the respective staple taper diagrams, determined by means of the SAWTRI Automatic Staple Length/Strength Tester, were highly correlated ($R = 0.97$).

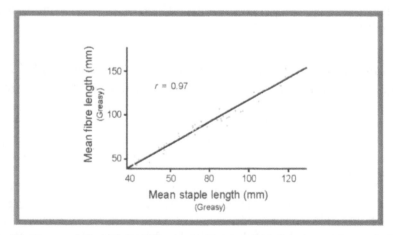

Figure 20. The linear relationship between mean fibre length and mean staple length in greasy wool [581]. Reprinted from M.R. Bow, *Staple length*, in *Proceedings of the Seminar 'Staple Length and Staple Strength of Greasy Wool: Measurement and Effects of Worsted Processing'*, CSIRO Division of Textile Physics, Ryde, New South Wales, December 1979, with permission of CSIRO.

Snippet fibre length measurement, using image analysis [551], has also been utilised to measure fibre length distribution within a staple. Schlink et al. [582] found fibre length parameters derived from snippet length to be different to those derived from staple length measurements of fibre length, the former underestimating the incidence of short fibres due to the presence of discontinuous fibres. This was explained in terms of the events leading to such (discontinuous) fibres to occur within a staple [583–585] and the possibility of fibre breakage when removing single fibres from staples. The above differences could also explain the previously reported failure of snippet length variation to show a significant difference between staple strength lines [576,586,587]. Schlink et al. [582] obtained a value of 18.5% for the proportion of fibres shorter than the staple length, this number changing significantly with years [588]. The incidence of shed fibres, or shut down follicles, for Merinos reportedly ranges from 0% to approximately 40% in the fleece [589,590], follicle shutdown being associated with nutritional inadequacies [585,591], which may account for the higher number of short fibres [582]. Ryder [592] reported low rates of fibre shedding for Merinos, although this increased for sheep grazing on seasonal pastures [582].

2.4.4. Staple length and fibre length variation

On average, about 70% of the variation in mean fibre length within a flock (mob) is accounted for by the variation in staple length [36,568,593], with up to 80% of the total fibre length variability being due to the variation between the fibres within staples. Another 10% of the total variation can be attributed to the variation between staples within the fleece. The remainder (about 10%) is due to the variation between fleeces (Figure 21) [593]. On the basis of the above, the classer is only able to remove about one-twentieth of the total variation present in fibre length within a flock [578]. Rottenbury et al. [594] have also concluded that within a mob the between-fleece component is less important than the length variability found within the fleece. It was confirmed that only a small portion of the variability found between individual fleeces is reduced by current classing procedures.

2.4.5. Top and sliver fibre length

It is important to note that considerable fibre breakage takes place from the greasy wool to the top stage, most of this occurring during carding, and generally increasing as the wool becomes longer, finer and weaker. Turpie [595] found the following relationship (Figure 22) between single fibre length in the greasy wool (X) and the mean fibre length in the top (Y):

$$Y = 0.586\,X + 14.4. \tag{22}$$

McMahon [596] estimated that fibre length variation in raw wool accounted for only 20% of the variation of the fibre length within a top, the rest of the variability being due to blending lots of different lengths, and fibre breakage, particularly during carding (see Table 9) [596].

Fibre breakage during processing increases significantly the fibre length variation [33, 36,560,593,597–601], in some cases more than doubling it [36,228], up to 50% of fibres being broken during processing from raw wool to top [602], with tender fibres more likely to break [603]. Few fibres are broken from the top to the knitted or woven fabric stage [604]. Various different fibre length distributions are experienced in tops [5,605,606],

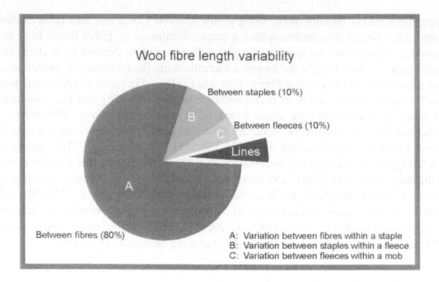

Figure 21. Components of wool fibre length variability [593]. Reprinted from J. Fowler, *Components of fibre length variability in greasy wool*, 4th ed., Australian woolclassing, 1990, p. 64 and The Indian Text J. 42 (1990), with permission of Samuel Joseph.

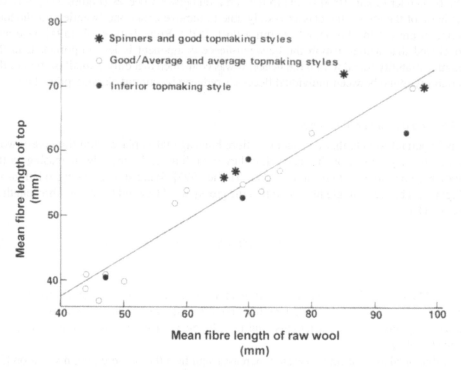

Figure 22. Relationship between mean fibre length of raw wool and top [595]. Reprinted, with permission, from D.W.F. Turpie, *The processing characteristics of SA wools, Part VI: The influence of style, length and class description on the processing performance up to combing of mixtures of South African wools*, SAWTRI Technical Report No. 250, June 1975.

Table 9. Increase in length variance from raw greasy wool to wool tops [596]. Reprinted from P.R. McMahon, Wool Technol. Sheep Breed. 23(1) (1976) pp. 8–10, with permission of IWTO.

Unit measurement	Source of variation	Length variability in fibre assemblies for wool tops	
		Coefficient of variation (%)	Percentage contribution to variability of tops (%)
Fibre	Single staples	10–20	10
Staple	Unclassed flocks	9–14	8
Staple	Shearer's contribution		Say 2
	Total for raw wool		20
	Blending and fibre breakage in carding and combing		80
Fibre	Tops	40–45	100

Figure 23. Almeter 100 (http://www.grupomarteam.com/archivos/Almeter100TS.pdf), reprinted with permission of IWTO.

this depending on various factors, such as blending, position of break and processing conditions.

For trading and technical purposes, fibre length is widely, and accurately, measured once the wool has been converted into sliver form, more generally into top. At present, the measurement of the fibre length characteristics of wool in sliver form, notably tops, is most commonly done on instruments based on the general principles of the Almeter [607] (IWTO-17–04), which could change with the development of new instruments, such as the OFDA4000. Couchman and Turk [607] reviewed the development and other issues relating to the Almeter and its measurement of top fibre length characteristics, and Couchman and Holmes [608] reporting on equivalence round trials for the Almeter AL2000.

The Almeter [607,609,610] (e.g. Almeter 100 and Almeter AL2000 (Figure 23) [611]), using capacitance, measures the length distribution of a fibre beard prepared by means

Figure 24. OFDA4000 (http://www.hornik.cc/ofda_4000.php), reprinted with permission of Mark Brims.

of a fibre preparer, which produces a 'length-biased' sample, the fibre preparation and measurement being automatic. A fibre diagram is produced and all the required parameters are computed by the software. The mean fibre length is usually expressed in terms of Hauteur (the average of the cross-section/linear density and capacitance-biased distribution of fibre length in wool top as measured by an Almeter [552,553,612]) or sometimes in terms of Barbé (where Barbé = Hauteur $(1 + V^2)$, with V = fractional coefficient of variation of Hauteur), Barbé being referred to as a weight-biased distribution. The coefficient of variation of Hauteur, CV_H, which generally varies from 40 to 60%, and the lengths of short and long fibres are also measured at the same time. The latter are usually expressed in terms of the percentage of fibres shorter than 25 mm (1"), and the length of the 5% longest fibres (tail), respectively. The length corresponding to the longest 5% of the fibres in the Almeter length distribution graph (Hauteur diagram) of a top is referred to as the Tail Length (Tail) and the percentage of fibres shorter than 25 mm is referred to as the Short Fibre Content (Sh25). Although the top fibre length distribution is often represented by CVH, Hansford [534] considered this a very crude measure of the fibre length distribution.

The correlation between mean fibre diameter, mean fibre length and short fibre content for commercial tops should be noted, Ogden [409] finding the following broad relationship for Merino and cross-bred tops:

$$H = 0.20D + 2.41, \tag{23}$$

where

1. H (in mm) is the mean fibre length (Hauteur).
2. D (in μm) is the air-flow mean fibre diameter.

Bernardin and Grignet [613] discussed the measurement of fibre length distribution in raw wool and fine woollen slubbings using an Almeter and the LAC technique.

The OFDA4000 (Figure 24), based on the OFDA2000 core technology, was unveiled at the IWTO in 2000 and introduced in 2002 [225,329,553,612,614–619]. It provides the same information as the OFDA2000, using automated microscope and image analysis techniques, to measure top fibre length and diameter distribution characteristics, such as Hauteur, CV_H, mean fibre diameter, CV_D as well as the length–diameter profile and fibre curvature in a single operation. It uses a small beard preparer to prepare the specimen for testing, and also measures length and diameter profiles along the beard. It measures fibre length and length distribution by optical scanning by means of a mathematical model, calculating the 'capacitance distribution' to produce values that approximate Hauteur- and Barbé capacitance-based distributions [553]. It produces a number weighted length, the fibre length distribution being the same as the Hauteur distribution when diameter does not change along the beard [224], although a significant fibre diameter profile along the beard has been observed. Fish et al. [618,620] compared the OFDA4000 results for diameter with those of the Sirolan-Laserscan™ system. The OFDA2000 profile measurements on the greasy wool were found to survive into the top as measured by the OFDA4000. It was suggested [224] that, with the addition of a second video camera, the OFDA4000 should be able to measure neps, VM and dark fibre.

The OFDA4000 can operate in 'native' mode, where it measures slivers for both length and diameter distribution, and in '70 mm' or 'snippet' mode, where it emulates an OFDA100 and measures 2 mm snippets, spread on 70 mm glass slides, for diameter distribution only [614].

An introductory paper on the OFDA4000 was presented at the IWTO meeting in Barcelona, a draft test method being proposed in 2005 [619]. In November 2005, the IWTO Regulations Committee approved the issue of Draft Test Method (DTM-62-05), covering the use of the OFDA4000 for the measurement of the length and diameter properties of sliver and tops. Baxter and Brims [614] reported on the performance of the OFDA4000 in three mill laboratories, which confirmed that under commercial conditions the instruments performed in accordance with the statistics given in DTM-62. Balasingam et al. [621] compared the OFDA4000 length results obtained on over 900 tops with those of the Almeter.

2.4.6. Fibre and staple strength

2.4.6.1. Introduction. One of the major consequences of wool processing is fibre breakage and reduction in mean fibre length, fibre strength playing an important role in worsted processing performance [622–624], in terms of fibre breakage during processing, particularly during carding [216], and in the strength of the yarn and fabric. Staple strength is second in importance to diameter in determining the price of greasy wool in Australia [1,9,581]. Therefore it is hardly surprising that considerable work has been directed towards obtaining a reliable measure of wool fibre strength, both rapidly and cost-effectively. It is a well-known fact that a strong, and virtually linear, relationship exists between single fibre breaking strength and fibre cross-sectional area (linear density or fineness), and that fibre cross-sectional area (or diameter), particularly its variation along the fibre, plays an important role in fibre breakage during processing [625–627]. As could be expected from a priori considerations, several research workers [103,440,453,455,628,629] found that finer fibres, which are weaker in terms of absolute strength (i.e. force to break), generally break more easily during processing and produce more noil [628,630]. Processing conditions also affect finer fibres more than coarser fibres [631]. A number of papers [632–634] have reported on the measurement of the tensile properties of wool fibre bundles and staples using various techniques and principles, Onions et al. [632] making reference to many studies carried out in this field.

Wool fibre strength can be measured and characterised in various ways. Essentially three main methods for determining wool fibre strength have been developed over the years, which are as follows:

- Single fibre strength measurement, which is very time-consuming and costly.
- Bundle fibre strength (IWTO-32–04), more suitable and used for tops and slivers.
- Staple strength (IWTO-30–98, amended and issued in 2007), widely used in the testing of greasy wool for various purposes, notably for certification and trading.

Each of the above methods finds application in specific circumstances. For example, the single fibre and bundle [632,633] strength methods are mainly used in laboratories for research and quality control purposes. Staple strength tests, on the other hand, are applied widely in the objective measurement of greasy wool (additional measurement programme), being more rapid and easily automated than the other two methods and also well related to worsted processing performance. Nevertheless, it is important to note that staple strength is a complex and compound measure and function of various factors, such as intrinsic fibre strength, fibre diameter variation (i.e. thin places), fibre length variation in the staple, fibre crimp variation, fibre alignment, fibre tension etc.

For commercial consignments and at a constant mean fibre diameter, bundle tenacity (BT) does not vary by more than ±10% [10], assuming no chemical treatment is involved. Yang et al. [635] gave the following equation for obtaining a very approximate measure of BT of tops, but it only accounts for about 45% of the observed variation.

$$BT = 0.54 - 0.21D + 0.05H + 0.0499SS. \tag{24}$$

At the 1996 ITMA, the CSIRO exhibited its Sirolan-Tensor for measuring wool fibre bundle strength [636], bundle strength always being lower than single fibre strength [637]. Yang et al. [638] described the Sirolan-Tensor for measuring wool fibre bundle strength in tops, stating that yarn tenacity varies in proportion to fibre bundle tenacity, and that a 9% decrease in bundle tenacity roughly doubled the spinning ends-down in top-dyeing damage trials. Fibre damage due to dyeing can increase spinning breaks [639,640], notably when fibre extension falls below 28–30% [639], with normal top dyeing having little effect [641–645], a bundle tenacity of around 10 cN/tex (100 N/ktex) is regarded [646] as the lower limit for dyed tops. Jerke et al. [647] reported on a staining test, which provided a measure of fibre damage and correlated with spinning performance.

Right at the outset it should be noted that the wool fibre strength (in absolute terms, i.e. uncorrected for fibre cross-section) in practice is largely determined by fibre diameter squared (i.e. by the fibre cross-sectional area), or more often by the lowest diameter (smallest cross-section, i.e. thinnest place) along the length of the fibre. Both fibre and staple strength are therefore generally determined by the cross-sectional area of the fibre(s), more specifically at the point of rupture, which is generally the place of minimum cross-section (i.e. thinnest place) [648–653] together with the tensile properties (i.e. intrinsic strength) of the fibre structure and material (alpha-keratin) per sé [364,654].

James [655] recommended the use of the minimum cross-section of the fibre for calculating stress, while Mason [656] suggested that fibres break at their weak places (natural flaws), which may not necessarily be places of minimum cross-section (i.e. thin places).

The term 'intrinsic strength' has been applied (defined) in different ways. For example

(i) the breaking force divided by the average fibre cross-section (diameter), this more generally being referred to as tenacity;
(ii) the breaking force divided by the minimum fibre cross-section (diameter);
(iii) the breaking force divided by the cross-section at the point of break, also referred to as specific stress at break.

Of the above three definitions, (iii) is the correct one, assuming that the cross-section is accurately measured, and is the representative of the cross-section of the undistorted fibre at the point of rupture prior to rupture.

Unless otherwise stated, wool fibre and staple strength will hereafter refer to the force to break divided by the *average* fibre or staple cross-section.

A great deal of published information exists on the causes of wool tenderness (i.e. reduced fibre strength or weakness) [657], these being summarised by Bingham et al. [658] and Hunter et al. [659]. Ryder [660] distinguished between 'tender' wools, which are weak along the entire fibre (i.e. staple) and will break at any point along the staple (for example, because of attacks from micro-organisms or a copper-deficient diet) and wools having a 'break' (tenderness or weakness), which occurs at a specific position along the staple. The latter is far more common than the former and is nearly always associated with a decrease in fibre diameter and can often be observed visually in staples because of the reduction in the diameter of all the fibres in the staple at that point [661,662]. This reduction in diameter usually has its origin in environmental factors, such as severe nutritional deficiencies, lambing stress, disease, parasites and seasonal (climatic) changes [659]. The Ryder *nomenclature* or terminology (distinction) is not widely used today, with the term 'tender' wool more often than not being applied to wool with a localised weakness (mostly thinner sections) [663].

In practice, a decrease in wool fibre strength can be caused by environmental factors experienced during fibre growth, which can either cause a decrease in the fibre diameter/cross section, or a decrease in fibre 'intrinsic strength' (e.g. due to weathering [664], copper deficiency, micro-organism attack etc.) or both. The former is by far the most common and troublesome. After the wool has been shorn, chemical treatments, such as scouring, dyeing [641–645] and shrink-proofing, excessive heat (e.g. over-drying/heating) and storage under hot and humid conditions (resulting in microbiological attack and breakdown), can also cause a decrease in fibre strength (i.e. in the intrinsic fibre strength). The effect of various farm management and related issues, such as, for example, nutrition, lambing and shearing date [27,534], which can affect staple diameter profile, strength (tenderness) and position of break, have been investigated [534,576,589,657,665–670] and also the processing consequences thereof [671,672]. As an example, pre-lamb shearing of Romney ewes generally produced sounder (higher staple strength) wools (up to two-and-a-half times stronger [673]) than post-lamb shearing, essentially due to the tender portion (lambing break) in the staple occurring close to the fibre end and not falling within the test length (i.e. within the portion or segment of the staple being subjected to the tensile stress). Such wool also produced superior processing performance [671], including higher yield, less noil during Noble combing as well as a longer top and stronger yarn. Scrivener and Vizard [670] reported no effect of Merino ewe age on staple strength. Disease or blowfly strike can also greatly reduce staple strength, for example from 40 N/ktex to as low as 5 N/ktex [11,534,675], which in turn can substantially reduce Hauteur. Nolan [11] reported that the stress of lactation can reduce staple strength by as much as 50% (for grazing dry, naturally non-supplemented pasture) but much less ($\approx$15%) under normal seasonal conditions, raising twin lambs resulting in a further reduction of between 4.4 N/ktex and 12 N/ktex. Disarranged or cotted fleeces display a significant decrease in staple strength (good top-making 36 N/ktex vs. cotts 28 N/ktex) [11], it being noted that the position of break often occurs at the change of the season [11].

A number of papers [535,648–650,674,676–678] have been published on aspects relating to fibre tenderness and strength, including intrinsic fibre strength.

Certain studies [677,679–681] indicate that the intrinsic strength or specific stress at break (i.e. breaking force divided by the fibre cross-sectional area at the point of rupture) of untreated wool normally does not vary widely; one of the few exceptions being severely weathered wool and steely wool, the latter arising due to copper deficiency in the diet of the sheep [680,682,683]. The average intrinsic single fibre strength or strength corrected for the cross-section at the place at which the fibre breaks, mostly the thinnest place, appears to be of the order of 210 N/ktex (21 cN/tex) [659]. Nevertheless, evidence has been presented that the intrinsic fibre strength (specific stress at break) can vary for different sheep [8,684,685] by 25 to 60% [685,686], and between different sheep breeds [8,650,652,654,661,684,687–689], and is lower for 'tender' than for 'sound' fibre segments [678]. In the latter case it is correlated with the half-cystine content [678], which is a measure of the cross-link density in the amorphous phase, implying that this phase is important in determining the intrinsic fibre strength. For sale lots, intrinsic strength can vary by at least 10% [8,684]. It should be noted, however, that not in all cases the results and conclusions were based upon the minimum cross-section or upon the point of break. Bundle tenacity generally tends to increase with mean fibre diameter [451,646].

As expected, there is not necessarily a correspondence between intrinsic strength and staple strength, because the latter is largely dependent upon diameter variation along the fibre, whereas the former is largely independent of fibre diameter variation along the fibre. Fibre diameter and length variations within the staple, notably the former, have the greatest effect on staple strength [690], which explains some 84% of variation in staple strength [582], with fibre crimp, weathering damage and shedding also having some effect. Staple strength is a measure of the average strength of the weakest places in single fibres modified by their collective behaviour normalised by the average, as opposed to the minimum, linear density of the staple [575], increasing with a decrease in gauge length [550]. The maximum staple strength is around 150 N/ktex [677,681]. Turpie et al. [544] suggested testing staple strength at two gauge lengths, the one at the maximum possible and the other at 20 mm in order to distinguish between wools of different 'intrinsic strengths'. Nevertheless, such a test can only provide a rough estimate of intrinsic strength. Bundle strength is not highly correlated with staple strength [8,684,691,692] unless the latter provides a measure of intrinsic fibre strength.

Clearly, the ability of the fibres within a staple to equally and continuously share the force (load) applied to the clamped staple ends during staple strength testing will play a crucial role in determining the maximum load which the staple can bear (i.e. the staple strength). Furthermore, weak places along individual fibres will determine the ability of each fibre to withstand forces and extension applied to the staple as a whole, so too with 'discontinuous' fibres, i.e. those clamped only at one end, or not at all, but still used in deriving the staple linear density (ktex) or cross-section. Due to variability in fibre alignment, crimped length, tension, extension and cross-section within a staple, some fibres will break before the others when a staple is subjected to tensile forces. Hence, the peak force recorded for such a fibre bundle will always be less than the sum of the individual fibre breaking strengths [690]. In a wool staple, the fibres have unequal crimped lengths and at a certain staple extension, the relatively short fibres, which are clamped at both the ends (i.e. in both the jaws), or the ones with less crimp, may be strained, whereas the longer or more highly crimped fibres clamped within the clamps may not yet be uncrimped [690]. De Jong et al. [690] found that the peak force required to rupture the staple was a slightly better indicator of fibre length in the top than the energy per unit mass. Fibre length variation in the staple affects staple strength by influencing the peak force but not the work to break [690]. The effect of fibre length variation in the staple was modelled theoretically and verified experimentally in certain studies [582], but not in others [586,587,693].

The role of along-fibre variation in cross-section (diameter) in determining both fibre and staple strength has been increasingly investigated and recognised. Various papers [231,531,653,694–696] have dealt with the relationship between fibre diameter variation along and between fibres, particularly the geometrical thin places [697], and fibre and staple strength, the variation in diameter within a fibre can be as large as that between fibres [231]. Both the fibre and staple strength are correlated negatively (inversely related) with variability in diameter (e.g. SD) along the fibre [23,179,231,481,520,524,531,693,698–700].

Hansford and Kennedy [523] had demonstrated the effect of minimum fibre diameter and the rate of change in fibre diameter along a staple on the staple strength, finding that the rate of change of diameter along a staple accounted better for variation in staple strength than did mean fibre diameter, minimum fibre diameter or the range in fibre diameter. For Romney sheep, the specific stress at break (i.e. strength corrected for the fibre cross-sectional area at the point of break) was generally not related to the fibre cross-sectional area, while more elliptical fibres tended to have a lower specific stress at break, nutritional treatment appearing to have little effect on the specific stress at break [701].

Wang et al. [702] verified that the strength and strain distributions of a single wool fibre fitted the Weibull distribution. Zhang and Wang [637] concluded that variations in the fibre breaking force (CV strength) of scoured and processed wool are mostly determined by the variation in fibre diameter, which is as follows:

$$CVstrength = 100 \sqrt{\left\{ 1 + \left(\frac{CVdiameter}{100} \right)^2 \right\}^4 - 1}. \qquad (25)$$

Wang et al. [703] similarly related CV of strength to CV_D for Mohair and Cashmere.

Diameter distribution, also that of the minimum diameter, generally follows a log-normal distribution [191–197,637] and there is an approximate 'linear relationship' between fibre breaking strength (force) and diameter squared [651,704,705] (also minimum diameter squared [652]) for dyed tops [637]. It therefore follows that the distribution of fibre breaking strength, as opposed to tenacity, should also follow a log-normal distribution [197,706], also after combing.

As most fibre breaks occur at the thinnest place, it follows that [637,707,708]

$$CV_{SFS} = \sqrt{\left[1 + (CV_{mfD})^2 \right]^4 - 1}, \qquad (26)$$

where

CV_{SFS} = fibre diameter variability of single fibre strength,
CV_{mfD} is the coefficient of variation of the minimum fibre diameter.

This better fitted the data. Two assumptions are also made in this respect [637], namely that there is a linear relationship between breaking force and the square of the minimum fibre diameter, and that the latter follows a lognormal distribution. Wang [709] applying the first assumption and a different approach, derived and verified experimentally the following simpler relationship on scoured wool:

$$CV_{BF} \approx 2CV_{mFD}. \tag{27}$$

Zhang and Wang [637] found that both predictive equations underestimated the actual values, with the first equation giving values closer to the actual ones, the discrepancies increasing as gauge length decreases. They demonstrated that even after top making and dyeing, breaking force variation could still be predicted from the variation in minimum fibre diameter, the results also indicating that most processed fibres break at their thinnest position.

2.4.6.2. Measurement of staple strength. For routine objective measurement and trading purposes, the measurement of staple strength offers maximum advantages, and considerable research effort has been directed [451,550,710] towards this end, with a high-speed automatic staple strength (and length) tester, which can also provide information on the position of break and 'tenderness' being the ultimate goal. The development, operating principles, evaluation and application of various staple strength testers developed over the years have been described, including the following:

- Simple portable tensile testers [711,712].
- The Agritest Staple Breaker Model 2 (SB2) [713–717].
- The UNSW PERSEUS staple length and strength machine [1,718].
- The ATLAS [714,719–721].
- The SAWTRI Automatic Staple Length and Strength Tester [544,562,563,722,723].

Traditionally, staple strength was subjectively measured manually (the flick test), but there were obvious limitations in this method in terms of the rapid and reliable estimation of staple strength, it generally being reasonably effective only for staple strengths below about 30 N/ktex – the level below which Australian Merino wools are generally discounted [724]. Ross [451,725] found a fair correlation (0.65) between instrument and hand-measured staple strength for 721 samples of Romney cross-bred wool, staple strength varying over the body of Romney ewes, which being highest in the neck and more variable for tender (lambing break) wools [550]. The flick test classifies wool as 'sound' (>30 N/ktex), 'part tender' (30 to 25 N/ktex), 'tender' (24 to 18 N/ktex) and rotten (<17 N/ktex).

The disadvantages associated with the subjective ('flick') test and other instrument testing led in the 1980s to the development of two instruments (see also Section 2.4.2), which could automatically measure staple strength (and simultaneously also staple length), namely the SAWTRI Automatic Staple Length and Strength Tester developed by Turpie and Cizek at SAWTRI (CSIR) in South Africa [544,562,563,722,723] and the ATLAS developed by the CSIRO [719] in Australia after early prototypes [720,721]. Both instruments enabled staple strength, staple length and position of break to be measured rapidly, automatically and reliably. Thompson et al. [726] discussed the design and operating principles of the ATLAS. The AWTA automated the staple sampling system for staple length and strength testing in their Mechanical Tuft Sampler (MTS), and in 1984 they chose the CSIRO ATLAS for their testing purposes. At present, the ATLAS is the only fully automated instrument, produced and used commercially.

Details of the Agritest Stable Breaker SB2 (Figure 25) semi-automatic staple length and strength tester were presented to the IWTO in 1997 [715–717]. It measures the thickness and the strength (force to break) of the staple. Baxter [727] reported that short staples

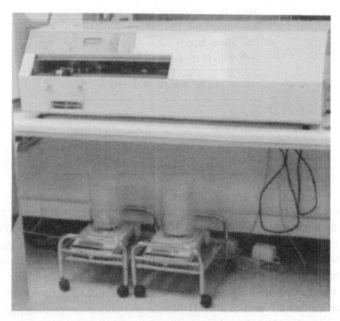

Figure 25. Agritest staple breaker model 2 (sanangelo.tamu.edu), reprinted with permission of IWTO.

were included in the Agritest Staple Breaker Model 2 (SB2) measurement procedures in accordance with IWTO-30 requirements, also suggesting that amendment be made to Appendix A of IWTO-30 to clarify these procedures. He discussed the effects of small numbers of short staples (shorter than 50 mm), such as second cuts, on staple length and its CV, concluding that their effect was most pronounced on the CV.

Vizard et al. [714] found good agreement between the ATLAS and Staplebreaker for Merino sheep mid-side samples, with the repeatability of the ATLAS being 4.13 N/ktex and that of the Staplebreaker 3.79 N/ktex.

It has been concluded [728] that, for the purpose of objective measurement, the most suitable parameter for indicating Merino wool staple strength is the normalised peak force to rupture, measured by using the entire length of the staple; this being the case for the ATLAS. Mid-side samples are appropriate for staple strength testing [657].

The instrument accepts over 60 staples on its feed belt and delivers them individually to the length-measuring section. Each staple is then conveyed past a bank of photocells, where its length is measured as the distance travelled by the staple while the light to the photocells is interrupted. The staple is then picked up by a pair of vertical belts and is fed to the jaws of the strength section. The staple is gripped at each end and stretched until it breaks whilst measuring the maximum applied force. The broken pieces are blown down to two balances where their masses are determined [170]. On the ATLAS, at least 60 staples are measured per lot. The relative weight of the two broken ends (pieces) are used to calculate the position of break (distance from tip, as a percentage of the staple length) [729], a correction being applied to allow for the VM in the tip, with the proportion of staples breaking in the tip, middle and base (root) regions being expressed as percentages. The test procedure is based upon IWTO-38–99 for grab sampling, IWTO-7–00 for sub-sampling staples from the grab samples and IWTO-30–98 for the determination of staple length and strength [730].

Table 10. Rating of staple strength [9]. Reprinted from I.M. Rogan, Wool Technol. Sheep Breed. 43(4) (1995) pp. 295–306, with permission of IWTO.

Staple strength description	
Sound	>30 N/ktex
Part tender	25–30 N/ktex
Tender	18–24 N/ktex
Rotten	<18 N/ktex

Fernandez and Marler [731] recommended that the minimum conditioning time for staples, for staple length and strength testing (IWTO-30–98), be reduced from 8 to 4 hours.

Commercially, the threshold between tender and sound wool is at a staple strength of around 25 N/ktex [12]. In practice, staple strength varies roughly between about 10 N/ktex (extremely weak or tender wool) and 60 N/tex (extremely strong or sound wool). A staple strength of below 25 N/ktex is regarded as 'usually tender', between 25 N/ktex and 30 N/ktex as 'tender/sound' and that of 31 N/ktex and above as 'increasingly sound' [730], and wool stronger than 30 N/ktex is generally classed as 'sound' [732]. Rogan [9] has given the following table (Table 10) for staple strength:

The average staple strength of Australian wool is around 35 N/ktex, the average varying little from year to year [9], although it can vary at a within-flock level from season to season and year to year [733], region to region [3,9,17,734,735] and between blood lines [733]. The average staple strength of New Zealand wool is 41 N/ktex [736]. Staple length CV of 12% or less is classed as excellent uniformity, that between 13% and 20% as good to average and of 21% or more as increasingly mixed [730]. The average staple strength of the weakest 25% (±15 staples) of the staples tested is also calculated and given [737] as SS25, a better than average SS25 is when it lies within 10 N/ktex from the overall average. Fine wools do not necessarily have lower staple strength, provided their along-fibre diameters do not vary unduly [724].

The following table (Table 11) gives rating staple strength in practice.

The SAWTRI instrument, which is no longer manufactured, measured mechanically the cross-sectional profile of greasy staples [544,562,563,722,723] by means of a tongue and grooved roller sensing system, from which the staple length and staple cross-sectional profile, including the presence of a 'break' (reduced or 'minimum' staple cross-section)

Table 11. Proposed rating for staple strength*.

Rating	Staple strength (N/ktex)
Extremely strong	>60
Very strong	51–60
Above average (strong)	41–50
Average (sound)	31–40
Weak (tender)	21–30
Very weak (very tender)	<20

Note: *Based upon average staple cross-section.

in the staple, were derived, after which the staple was automatically gripped at pre-set distances from its two ends and broken. The staple tenacity, i.e. cross-sectional corrected strength of the staple, was then automatically calculated. An appropriate correction factor for grease content was also applied. The POB, derived from the staple profile, corrected for the varying degree of contaminants and grease along the staple from root to tip, was assumed to take place at the position along the staple where the staple cross-section (profile) was at its minimum, this being mostly the case. If no such place was detected, which is normally the case for 'sound' or 'strong' wools, the absence of a break in the staple was assumed, and the POB was accepted to occur at random along the length of the staple, averaging out at 50%. A high correlation between the actual and derived POB was found [544], reaching $R^2 = 0.96$ when there was a pronounced break or tenderness in the staple.

Various other studies [583] on fibre and staple strength have been carried out. Attempts have been also made to develop staple strength prediction models [738]. A system for measuring the extensibility (stretchiness) of staples was developed for integration into the ATLAS instrument [536].

Schlink et al. [582] presented a correlation matrix table, showing the correlation between various fibre length and diameter characteristics and staple strength, also deriving various relationships involving staple strength and fibre properties.

2.4.6.3. General conclusions.

From the various papers related to wool fibre and staple strength, the following main conclusions can be drawn:

- Fibre diameter (cross-sectional) variations are responsible for the bulk of variations in fibre and staple tenacity obtained in practice, where the cross-sectional area (diameter) used to calculate such tenacity is not that at the point of break or the minimum (thinnest place) within the test length.
- When a fibre or staple is extended, it mostly breaks at the thinnest place.
- When using the actual or minimum cross-sectional area to calculate single fibre tenacity, and/or when testing at a very short gauge length, the tenacity (specific breaking stress), is around 210 N/ktex, but can be somewhat different for sound and tender wools and from different sheep and sheep breeds.
- Staple strength and bundle tenacity tend not to be related.
- The bundle tenacity of commercial undyed wool tops is fairly constant at around 100 N/ktex, increasing slightly with increasing mean fibre diameter and decreasing with fibre crimp (or resistance to compression).
- Chemical treatments, such as shrink-proofing and top dyeing, under normal conditions can reduce fibre tenacity by up to 10%. It also reduces fibre extension, which is more sensitive to damage as a result of chemical treatment.

2.5. Colour

2.5.1. Introduction

Although the colour of wool fibres is important, the colour of greasy wool does not provide a reliable measure for the colour of the scoured (clean) wool [172,739], it being determined by the inherent colour of the wool as well as by that of the grease, dirt, suint and VM present. Pattison and Whiteley [154] highlighted the poor relationship between 'unscourable' colour subjectively assessed from samples of greasy wool and the colour measured in the sample of clean wool. The colour of wool, more specifically after commercial scouring (i.e. the non-scourable colour), plays an important role in determining its perceived quality and price [740–745], as well as its end-use, mainly because it is the best indicator of dyeing potential [172,746–749], more particularly the shades to which the wool can be dyed [747]. The presence of 'unscourable colour' in wool can prevent the wool from being dyed successfully to pale and pastel shades, and being applied in end-uses, where pale shades are fashionable or desirable; ideally the wool should be white and bright [743–746]. The colour of the clean or scoured wool is a good indicator of the colour of the top [750,751], the more so for average yellowness (Y–Z) than for brightness (Y), the top generally being slightly whiter, with brighter wools tending to be whiter (less yellow) [750]. Colour, when related to bacterial or weathering damage (i.e. fibre damage or weakening), can also affect processing, fibre breakage in particular, Rottenbury [752] having discussed the relationship between wool colour and processing.

Disregarding urine stains and pigmentation, it has been suggested that there are three main types of wool discolouration [753]:

- 'Butter' colour. Substances in the wool auto-oxidise to yellow/brown; thus all wool will discolour to some extent during storage. The most discoloured wool will turn yellow the fastest, most of this colour being scourable [753].
- Canary yellow (yolk stain).
- Bacterial stains.

Base colour is defined [754] as 'the inherent colour of either greasy or commercially scoured or carbonised wool after the removal of contaminants (e.g. grease, suint, dirt, VM) by laboratory cleaning processes'. *As is colour* is defined [754] as 'the colour of the sample of commercially scoured or carbonised wool as received at the laboratory'. 'As is colour' therefore refers to the colour of greasy wool after it has been scoured, carbonised, washed or solvent-degreased (termed raw wool), but contains variable quantities of VM and extraneous alkali-insoluble impurities, mineral matter, wool waxes, suint and moisture, it being demonstrated [755] that re-washing of scoured wool can improve its colour. The 'as is colour' provides the measure of efficiency of the scouring process, using the 'base' colour as a benchmark. The colour of greasy wool per sé (i.e. not cleaned) does not provide a reliable indication of either the 'base colour' (i.e. intrinsic) or the 'as is colour' of the scoured wool.

Various authors have reported on wool colour and the factors that affect the colour of scoured wool [744,755–761], including weathering [762]. It should be noted that wool colour may vary within the staples, for example from more yellow tips, because of their increased weathering, to whiter roots, which are better protected from weathering. Also, clumps or 'bands' of 'more yellow' wool can also occur, for example due to urine staining and fleece rot, which complicates representative sampling, and the accuracy of results. Johnston [136] referred to complications arising from measuring only the 'average colour' (average yellowness), as it neither reflects variations in colour (e.g. 10% discolouration), which could influence dyeing, but not the other colours that may affect greasy wool from time to time. Precision could also present a problem for both buyer and dyer.

Holt et al. [762] showed that wool from the sheep's back had a poorer colour than that from the flank due to increased weathering, and that rugging the sheep [763] and other treatments reduced weathering and improved colour. Removing the inferior colour fleece components, primarily the skirtings and belly wool, can effect some improvement in colour [53]. Holt et al. [762] reviewed and summarised the effects of weathering and weathering control measures on the colour of scoured wool. When wool is exposed to sunlight, water and other elements of nature during its growth, referred to as weathering, it gets damaged [764], mostly in the tip region. This manifests itself in differences in the dyeability (tippy dyeing) between the tip and other sections of the staple, increased processing losses, discolouration and deterioration in other fibre properties [762]. Weathering results in changes in the chemical composition of wool, which provides a measure of the degree of weathering [762]. Changes in the uptake of dyes (such as methylene blue and a mixture of acid dyes), primarily due to cuticle damage, are commonly used in practice to determine the degree of weathering [762,764–768], which can extend by as much as 30 mm from the staple tip of fine wool. Most, but not all, [766,768,769] of the weathered wool tips break off during carding and are removed during processing.

Heating, moisture, alkalinity and sunlight can all cause yellowing of wool. Wool colour is related to suint content, modified by temperature, humidity, rainfall and sheep breed. Environmental conditions, such as high humidity and temperature [770], which promote fungal and bacterial growth on the wool, play a dominant role in the development of colour in raw or greasy wool [771–773]. Bacterial stains (fleece rot), water stains, weather stains etc. are caused by colonies of bacteria which multiply in warm and damp conditions, such as at the base of the fleece (notably on the back of the sheep) when the wool is wet [753], often occurring in bands. Such stains are usually yellow, but may be green (second most common stain), brown or apricot. Some bacterial stains are scourable [753]. Severe yellow discolouration, sometimes referred to as 'canary yellowing', can also develop during wool growth, particularly in the regions of the belly and lower edges of the fleece, as well as during storage (notably under warm and damp conditions) and processing [740,771,774]. Canary yellow is associated with dampness, warmth and high alkaline suint content and is mainly present in the cuticular region of the fibre [740], and probably results from the Pseudomonas infection [740]. Most of the colour is scoured when a considerable amount of grease is present, but not when little grease is present.

The colour of greasy wool, and to a lesser extent scoured wool, can deteriorate significantly when it is stored under hot and humid conditions [739], which provides ideal opportunity for bacterial and fungal growth and their activity on the wool. Nevertheless, it is not always the wool colour per sé that changes during storage, but often the grease, which is oxidised and changes colour. Greasy Merino wool's colour changes relatively little during normal storage, but that of greasy cross-bred and half-bred wools can deteriorate, especially under high temperature and humidity conditions [775]. It has been reported that, even after years of storage of Merino wool in dumped form, little change in the clean wool colour occurred [750].

The above details are important considerations in the measurement of colour as the colour could change over a period of time, depending upon the state and type of the wool and storage conditions.

Laboratory methods have been developed to measure the propensity of wool to yellow [739,758], two factors, namely genetic and environmental, being involved in wool yellowing. Wilkinson described the method by which the susceptibility of greasy fleece wool to yellowing can be predicted.

The reasons for, and fundamental aspects of, colour and colour changes, notably due to photo-yellowing, have been discussed in depth by various authors [776–779]. Simpson [780] published a detailed review of the physics and chemistry of wool colour and yellowing, and so did Millington and co-workers [777,781]. Dyer et al. [779] found non-scourable yellow discolouration to be located predominantly in the cuticular region of the wool fibre. Wood et al. [783] found that the compounds responsible for non-scourable canary yellowness in New Zealand wool were N-formylkynurenine and kynurenine, together with dityrosine and other unidentified phenolic compounds; this confirmed the works done by other researchers on Merino wool. Brown [784] dealt with generic aspects of greasy wool colour for Merino sheep.

2.5.2. Colour measurement

In recent years, the measurement of wool colour has received increasing attention, and tests for the measurement of colour on greasy wool after a standard scouring operation as well as on wool slivers and tops have been developed. A number of papers [166,753,785–795] have discussed the measurement of wool colour and the implications thereof; Marler [785] discussed it in detail, including the sample preparation.

As the intrinsic or scoured colour of wool is very difficult to predict from the appearance of the greasy wool, the greasy wool has to be prepared (e.g. scoured or cleaned) in a standardised way before colour measurement is done. Care must be taken to ensure that drying procedures involving drying at 105° or higher do not change the colour of the wool [785], because drying at this temperature or higher can decrease the wool brightness and increases its yellowness over time [796].

The colour of New Zealand wool has always been an important attribute and it is therefore not surprising that WRONZ played a leading role, commencing in the late 1960s [176], in developing an objective test to measure the colour of clean and blended wool samples, i.e. 'base colour' [557,797,798]. A draft Australian Standard became available in 1976 and a New Zealand Standard Method of Colour Measurement of scoured wool in 1977 [172], which culminated in a New Zealand Standard (NZ S 8707) in 1984 for both scoured and greasy wools. An Australian standard method for the measurement of the average yellowness of cleaned greasy wool was published in May 1988 [753]. The Australian method provided brightness in terms of L* and yellowness in terms of b*, whereas the New Zealand method used Y and Y–Z, respectively. A joint Australian–/New Zealand working group drafted an internationally accepted IWTO test method under examination in 1988 [172], and after some changes, the draft test method (IWTO DTM-56) was granted full IWTO test method status in 1999 [799], with the latest [642] version (IWTO-56–03) containing standardised sample preparation and calibration procedures [800].

The measurement of the colour of raw wool is described in the test method IWTO-56–03 (issued after an amendment in 2007), and that of wool sliver in IWTO-35–03. It is aimed at providing a measure of the inherent (i.e. non-scourable) colour of raw wool, and is usually expressed in terms of the Yellowness Index (Y–Z) derived from the tri-stimulus values Y and Z. Y indicates the brightness (whiteness) of the wool, the higher the value of Y the better the brightness of the wool. The lower the yellowness (Y–Z) the better, skirting wools generally being yellower than fleece wools [750]. Finer Merino bloodlines in Australia tend to have a better colour (not as yellow or stained) than coarser bloodlines [733,739]. Raw wool colour is commonly measured on the scoured and Shirley-analysed sample used for fibre diameter measurement.

The colour of wool is obtained by measuring the proportion of reflected light from its surface throughout the visible spectrum, but in practice the reflections in the red, green and blue regions are considered sufficient. Under the specified conditions of International Commission of Illumination (CIE) these are referred to as X (red), Y (green) and Z (blue). They can be used directly or transformed to provide information on specific aspects of colour, such as yellowness (Y–Z).

Colourimeters, spectro-colourimeters (Figure 26) and spectro-photometers, even NIR [801], have been used to measure wool colour. Initially, tri-stimulus colourimeters were

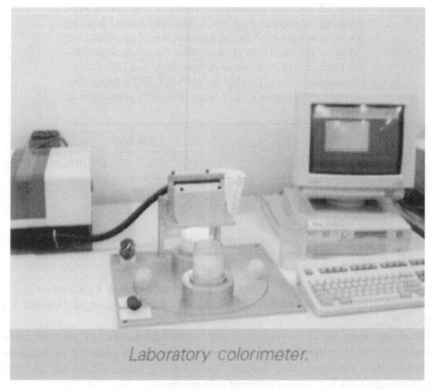

Laboratory colorimeter.

Figure 26. A laboratory colorimeter (www.awta.com.au/Documents/News%20Letters/2003–02-pdf/Feb_2003_News.pdf), reprinted with permission of AWTA Ltd.

used for measuring wool colour, but today more sophisticated colour spectrophotometers are generally used in which the light reflected from the wool is split into its component wavelengths by a reflection grating, each being sensed in turn by a detector. Computer software converts the different wavelength signals, received by the detector, into X, Y and Z values. Normally, sufficient measurements are taken to ensure that precision is better than ±1 unit for Y–Z [746]. The thoroughness with which the sample is prepared and cleaned can significantly affect the values obtained, as can the method of presentation, particularly the packing density of the sample in the cell, and the type and thickness of the window glass against which the sample is pressed etc. [746].

Initially wool colour was measured with instrument calibration being carried out by means of wool calibration samples and using the C illuminant and 2° observer angle, (the C/2° system). In 1964 the CIE introduced the 10° observer angle to replace the 2° observer angle, because the former was more realistic. For various reasons the C/2° system was subsequently replaced by the colour measurement using a 10° observer angle, ceramic tiles, instead of wool samples, for calibration and the D65 illuminant (D65/10°) system, the results obtained by means of the two systems being highly correlated but not the same, there also being no simple relationship (transform) between the 'old' Y–Z and the 'new' Y–Z values [799]. The IWTO in 1997 recommended that the D65/10° system should be used, the official change occurring in 2002, with the L* (brightness) corresponding to Y and a* corresponding to Y–Z (i.e. yellowness). Changing from illuminant C to illuminant D65 hardly changed the Y values, but significantly changed the Z values, and therefore the

Table 12. Colour (yellowness) values for Australian wool [754,785]. Reprinted from S.A.S. Douglas, Amendments to IWTO-56-99 'Method for the measurement of base and as is colour of raw wool', IWTO Raw Wool Group Appendix No. 03, Christchurch, New Zealand, April 2000 and J.W. Marler, Amended version of DTM-56 for raw wool, IWTO Report No. Appendix 02, Florence, Italy, May 1999, with permission of IWTO.

Colour description	Y–Z units	
	C/2°	D65/10°
Very white	−2	7
White	0	8
Creamy	3	11
Canary yellow stain	12	18

Y–Z (yellowness) values, a similar effect applying when changing the observer angle from 2° to 10°. Shifting from wool calibration to tile calibration also changed both the Y and Z values [799].

The change from the C/2° system and wool calibration to D65/10° and tile calibration resulted in a change in the obtained colour values, the results obtained by using the two different systems that are related linearly. The following values (Table 12) are given for Australian wool [754,785], using different calibrations:

Wood [774] provides the following table (Table 13) for New Zealand wool, based upon the C/2° system.

Colour transformation equations from wool calibration to tile calibration were given by Shepherd et al. [794], Crowe and Shepherd [802] arriving at the following empirical relationships between $C/2°_{(W)}$ and $D65/10°_{(T)}$ results on wool:

$$X_T = 1.026 \ X_W + 3.226, \tag{28}$$

$$Y_T = 1.094 \ Y_W + 0.953, \tag{29}$$

$$Z_T = 0.971 \ Z_W + 0.596. \tag{30}$$

The lightness or brightness of raw wool is measured in terms of the reflectance intensity and can be expressed as either the Y tri-stimulus value or whiteness index (W):

$$W = [(100 - 0.94Y)^2 + (2.84X - 2.35Z)^2].^{1/2} \tag{31}$$

Table 13. Typical colour values for New Zealand wool [774]. Reprinted from E. Wood, Wool Technol. Sheep Breed. 50(2) (2002) pp. 121–132, with permission of IWTO.

Wool type	Good colour Merino	Poor colour Merino	Good colour crossbred	Poor colour crossbred
Brightness (Y)	>67.5	<65.0	>61.0	<57.5
Yellowness (Y–Z)	<1.5	>2.0	<4.5	>6.5

The CIELAB L*, a* and b* system of coordinates gives another expression of colour, in which

$$L^* = 116(Y/100)^{1/3} - 16, \tag{32}$$

$$a^* = 500[(X/98.1)^{1/3} - (Y/100)^{1/3}], \tag{33}$$

$$b^* = 200[(Y/100)^{1/3} - (Z/118.2)^{1/3}], \tag{34}$$

where

L^* represents brightness,
a^* represents the green–red axis,
b^* represents the blue–yellow axis.

Qi et al. [803] developed a computer algorithm for measuring the colour of scoured animal fibres, using the Texas Agricultural Experiment Station's (TAES) Automatic Image Analysis System.

2.6. Coloured (dark) and medullated fibres

2.6.1. Introduction

Various papers [804–807] have addressed issues and problems related to the presence of dark (pigmented and stained) and medullated fibres in wool, Giebel [808] discussing the problems that cause a spinner. The terms 'dark' and 'coloured' fibres are generally interchangeable and cover both stained and pigmented fibres.

Coloured (dark) fibres in wool generally encompass both stained (e.g. urine or fungus stained) fibres and, more specifically, what is commonly referred to as 'black fibres', which are pigmented and could vary from brown to black, it being not so easy to distinguish between stained and pigmented fibres except under a microscope. Pigmented fibres contain discrete granules of colour (pigments), which are 2 μm in size [809] whereas the colour of urine-stained fibres is due to a fairly uniform staining throughout. The presence of such fibres in good quality 'white' wool, if detected, attracts a substantial discount in price [804,810–812].

Dark (coloured) and medullated fibres, within the ranges normally encountered in good quality Merino-type apparel (white) wools, have little effect on processing performance. Nevertheless, such fibres can present a serious contamination problem, representing an aesthetic or visual flaw (defect) in the final product, particularly in undyed or pastel shade products, causing the product to be downgraded as seconds or rejects, and possible claims.

Dark fibres. For good colour, good quality white Merino wool, a level of 100 dark fibres per kilogram of top, is often taken as the commercially acceptable limit [6,807,813]. On this basis, only one 'black' staple of 4000 fibres ($\approx$5g) in 10 fleeces can present a problem [809,814–816]. There are some 60×10^6 fibres on a sheep; therefore only 400 dark (black) fibres on the sheep can present a problem, this representing a level of one fibre in 150,000 or 0.0007%. For critical end-uses, e.g. pastel shades in piece-dyed fabrics or knitwear, a count of even 30 coloured fibres per kilogram can lead to rejection [112]. Rottenbury et al. [77] confirmed that only a small quantity of urine-stained fibres (which retain much of their colour after scouring [807]) or pigmented wool can cause very high dark fibre counts in tops.

A major source of dark fibre (DF) contamination in Australian adult Merino wool is from urine-stained fibres, with pigmented fibres mostly of secondary importance [817]. When the number of dark fibres was below or close to the commercial limit [818] of 100 dark fibres per kilogram, the proportions of pigmented and stained fibres tended to be almost the same, but when it exceeded this, the majority of such fibres were usually stained [817]. Exceptions to this were mainly found in sale lots from young sheep, where the pigmented fibre count was usually in excess of the average level of 23 DF/kg for sale lots of adult Merino wool [817].

Black or pigmented fibres in the wool of a sheep are generally determined genetically (due to the gene being present in the sheep itself) but could also arise from cross-contamination by animals having such fibres (or even hair). In practice, cross-breeding (i.e. genetics) and running flocks of white wool Merino sheep and sheep with black fibres together are mainly responsible for this type of contamination.

It has been reported [819] that the dark and medullated fibre levels in the top are generally lower than expected, as based upon the levels in core samples, only 40–50% of dark fibres present in the raw wool samples are detected in the top [820,821], possibly due to such fibres being shorter and therefore removed during processing. Fleet and Foulds [818] reported, however, that the number of pigmented fibres in tops processed on the CSIRO mini-scale plant was twice as high as those counted in the fleeces, ascribing this to fibre breakage during processing, removal of fleece sample staple tips, inferior conditions when viewing tops or differences between observers. Fleet et al. [822] reviewed the results of processing trials, and studied the relationship between pigmented fibre levels in the raw wool and those in the processed top. They found that the levels of inherent (genetic) pigmented fibres in the top were higher than those in the raw wool due to fibre breakage during processing, whereas the reverse was true for cross-contamination pigmented fibres, ascribed to the fact that such fibres were generally shorter and, therefore, preferentially removed during combing [822,823].

Medullated fibres. Merino wool fibres generally have a solid cross-section (cortex), but sometimes medullated fibres are present which are not solid, having a hollow (air-filled) central canal (medulla) or one partially filled by cell-remnants, running along their entire length or part of their length (heterogeneous). Medullated fibres are essentially genetically determined, but can, as in the case of pigmented fibres, also arise from cross-contamination. The more extreme forms of medullated fibres are classified as kemp fibres or as 'objectionable medullated fibres'. The presence of such visually identifiable (objectionable) medullated fibres results in a substantial discount in good quality wool products. Even a very low percentage (<0.1%) of such 'objectionable' fibres can present a serious quality problem, mainly because of their presence being regarded as a 'defect' or 'flaw' in the final product.

Balasingam [824] has recently prepared a detailed review on definitions and threshold values related to 'objectionable' medullated fibres in Merino wool. Such medullated or kemp fibres are generally coarser than the bulk fibres, and, obviously, also have a lower density than the unmedullated (i.e. solid fibres). The main problem associated with such fibres is their different appearance (light or milky white), noticeably after dyeing, resulting largely from differences in the light-reflecting properties of the medulla and the fact that less solid matter (protein) is available to take up dye. Such fibres, therefore, appear paler in shade (even undyed) after dyeing, or white (milky white) in the undyed form. They can, when present at very high levels, also affect handle (scratchiness) and processing performance and waste.

The ratio of the medulla diameter to the fibre diameter generally provides a measure of the severity of the effect and degree of medullation; the greater the ratio, the more severe the effect, and if it exceeds 0.6 (or 0.65), the fibre is commonly defined as kemp. The AWTA and AWI have indicated a med ratio of 0.8 in a fibre of 50 μm as an appropriate contamination threshold [821]. Nevertheless, the final answer has not yet been provided as to which property (properties) or characteristic(s) reliably differentiates between an 'objectionable' medullated fibre (i.e. one which is clearly visible to the naked eye) and a 'not/non-objectionable' (i.e. 'acceptable') medullated fibre.

Medullated fibres counted by projection microscope generally follow a Poisson distribution, with 95% confidence limits as follows [239]:

$$\text{Lower confidence limit} = m \left[1 - (1/9m) - 1.96\sqrt{(1/9m)} \right]^3 , \qquad (35)$$

$$\text{Upper confidence limit} = M \left[1 - (1/9M) - 1.96\sqrt{(1/9m)} \right]^3 , \qquad (36)$$

where

m = observed count,
$M = m + 1$,
(total number of fibres examined = 1000).

2.6.2. *Coloured and medullated fibre measurement and risk schemes*

Coloured and medullated fibres generally occur as very isolated events in Merino-type wools and are therefore extremely difficult to accurately measure, being like the proverbial 'needle in the haystack'. Traditionally, the number of such fibres in a sample was determined by time-consuming and tedious manual sorting and counting methods; for example, by physically picking out manually/tweezers and counting and weighing the fibres in a sample of certain mass, or by examining, measuring and classifying fibres under a microscope. Manual counting and classifying such fibres in a top under balanced illumination on a Toennissen-type instrument later followed, and was eventually automated in the form of Centexbel Optalyser. Lemaire et al. [825] described the automatic instrument (the Optalyser (OP-300)) based on advanced technologies in artificial vision and image analysis (CCD camera), developed by Centexbel to detect, classify and count cleanliness faults (neps, slubs, VM and coloured fibres) in tops. The Optalyser OP-300 has been approved as an IWTO test method (IWTO-55–99). It automatically detects, counts and classifies faults (neps, VM and dark fibres) in wool slivers and tops, using pre-determined threshold levels. Longree et al. [826] compared the Optalyser cleanliness fault results with those obtained by visual counting, and Longree and Delfosse [827] reporting on the ways of improving the reliability of cleanliness tests on wool tops. Fleet et al. [820] found a good correlation between dark fibre levels counted in fleece samples (using balanced light illumination – DTM-13–01) and those measured in the top by the Optalyser or the benzol top test, the levels in the top being 40 to 50% that in the core sample.

Various attempts, with varying degrees of success, have been made to develop a rapid objective test for measuring medullation, including the WRONZ Medullameter [828], the OFDA system [829] and the NIR [176,830–834]. Within this context it is important to distinguish between the measurement of medullated fibres and the measurement of the 'objectionable' or 'kemp' type of medullated fibres, because the two are by no means equivalent, and not necessarily highly correlated. For example, the NIR system can be used to measure total medullation fairly accurately, but not the level of objectionable medullated (kemp-type) fibres [176].

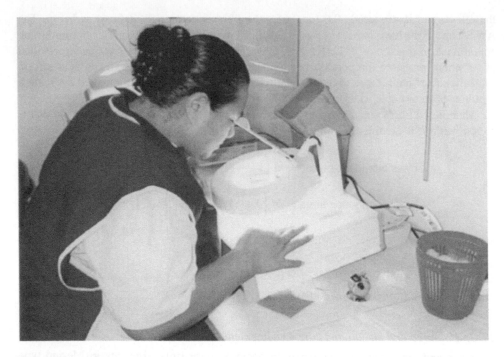

Figure 27. CSIRO dark fibre detector (http://www.awta.com.au/en/Home/Publications/ Newsletters/), reprinted with permission of CSIRO.

In a global survey conducted by the AWTA in 2004 [835], medullation was essentially measured on tops using a microscope; if the area under examination showed 50% or more medullation, the fibre was considered medullated.

An illumination attachment and software enhancement in the OFDA provides a second (dark field) illumination system, which highlights the medulla, alternating the two illumination systems and enabling fibre diameter and medullation to be measured simultaneously [557]. Turpie and co-workers [338,836] carried out extensive and definitive studies on the application of the OFDA100 system for measuring medullation and objectionable medullated fibres in Mohair, using the degree of opacity as the main criterion; their work also being applicable to wool.

Maher et al. [837] investigated the accuracy and precision of the OFDA-measured medullation content by volume, finding that the exponential transform gave the best precision and accuracy, and was potentially sufficient for commercial purposes. They also suggested further improvements to increase the accuracy.

The dark and medullated fibre pre-sale test developed by the CSIRO and introduced by the AWTA [838] involves immersing a 5 g sub-sample into benzyl alcohol, which has the same refractive index as the wool, in a sealed plastic bag, which makes normal white wool transparent while the coloured fibres (urine-stained and pigmented ones) appear dark against a white background and the medullated fibres appear white against a black background.

The CSIRO developed a Dark Fibre Detector (Figure 27) and 'Risk Scheme' [807, 817,830–832,839] (e.g. Sirolot™ , a clip lotting program) [813] in an attempt to more objectively and accurately quantify the presence and risk of such fibres, respectively. A

Dark and Medullated Risk Scheme [831,832] for Merino fleece and pieces wool was implemented in Australia on 1 July 2004 [833] (http://www.awta.com.au). Due to difficulty in sampling greasy wool for dark fibre measurement, the dark fibre risk scheme was developed [817,830], which concluded the following:

- Ewes presented a greater risk than wethers.
- Mulesing and crutching reduce stained fibres.
- The time lapse between crutching and shearing is important in stain formation.
- There is a relationship between isolated pigmented fibres and the age of the sheep.
- Blending wool from different wool categories increases the risk.
- Inspection of sheep for urine stain is difficult.
- Quality management reduces dark fibre risk.
- Certain sheep breeds present a greater risk than others.

The CSIRO also developed a dark fibre classifier to assist in classifying the darkness levels of individual fibres and supplements its dark fibre detector [807]. This classifier is basically a micro-photometer, which uses optical transmittance through a single fibre to ascertain its bulk reflectance properties [807]. The CSIRO Dark Fibre Detector, using balanced light illumination (IWTO-13—88), was identified [840] as the best of techniques (e.g. visual, image analysis, OFDA, NIR, UV fluorescence, flotation and CV of diameter) for identifying and quantifying dark and medullated fibres in greasy core samples (i.e. not slivers or tops). The CSIRO developed a colour scale composed of dyed wool fibres [809], going from white (labelled 0) to black (labelled 8). On this scale, a darkness level greater than 4 (i.e. 5 and more) was considered to be commercially problematic (unacceptable) [809]. Coloured fibres (CF) can be taken to comprise CSIRO colour reference levels 4 and 5, and brown/black fibres (BF) levels 6, 7 and 8 [835], with the total of these two categories representing the total DF content. It was found that fibres of darkness grade L5 (level 5 on the CSIRO reference scale [841]) could be detected in white fabrics [814], with more than 6 DF/kg visible in the fabric (i.e. 3 DF per garment) possibly leading to claims against the spinner [814]. Foulds et al. [814] concluded that 100 DF/kg, as measured with the CSIRO Dark Fibre Detector (IWTO Method), is the upper limit of acceptable levels of contamination for knitted structures, and that for woven fabrics probably less, with fabric colour having a significant positive influence on reducing the detectability of DF. There was, however, a wide range of detection efficiencies using different inspectors.

The introduction of exotic and meat sheep breeds in Australia increased the contamination by dark and medullated fibres, a risk factor of 6 being applied to categories where there has been direct contact with exotic sheep, such as Awassi, Damara, Dorper and Karakul breeds [833]. The development and application of the Dark and Medullated Fibre Risk Scheme [819,838,840,842–844] have been discussed, with Hansford reviewing the subject matter.

Delfosse and Grignet [845] have compared the optalyser results with those obtained by means of the CSIRO Dark Fibre Detector.

Hansford and Swan [835] reported on the findings of the survey of dark and medullated fibres involving some 689 wool consignments processed in 14 mills during 2004. They concluded that there had been no detectable increase in such fibre contamination during the period 1999 to 2004 and that Australian wool suffered relatively less from such contaminating fibres compared to wools from other countries; they further concluded that the use of Dark and Medullated Fibre Risk Scheme would assist buyers when sourcing wools for sensitive end-users.

Burbidge et al. [846] reported on dark fibre contamination in tops from consignments of Australian wools.

Chudleigh [847] has described a method of detecting coloured contaminants in webs of white wool, using an integrating sphere.

The CSIRO, the AWI and the AWTA have jointly developed an automated imaging system for the detection of dark and medullated fibres in tops, but at present it can only count dark fibres (AWTA Ltd newsletter, April 2007).

2.7. Fibre crimp

2.7.1. Introduction

Crimp is an important fibre property affecting fibre cohesion, drafting, bulkiness, processing performance, spinning, yarn quality, fabric handle, comfort and bulk etc. Crimpiness is mostly quantified in terms of crimp (wave) frequency, rather than in terms of wave or crimp amplitude. Higher crimp tends to be associated with less entanglement during scouring, lower yarn hairiness and neps and fabric felting shrinkage, but higher noil, yarn irregularity and thin and thick places and spinning end breaks and lower the yarn strength. Higher crimp wools tend to be stiffer than lower crimp wools, which impacts their handling properties [848].

Apparel wools, notably Merino, generally exhibit a natural crimp (sinusoidal or helical form), largely due to their bilateral (ortho-cortex/para-cortex) structure; higher crimp being associated with higher levels of para-cortex and sulphur [849] (mainly in the high-sulphur proteins of the wool [850]) and is related to the shape of the skin follicle [851,852], which varies according to position on the sheep. Staple crimp increases with decreasing nutritional levels and tends to be negatively correlated with the staple strength [852]. Differences in the fibre properties, processing behaviour and yarn and fabric properties associated with differences in crimp, could therefore be due to differences in either the fibre's physical or chemical structure or both or even due to differences in the fibre extent. For example, Whiteley [853] maintained that differences in 'fibre substance', possibly manifesting in fibre creep behaviour, also affected the handle and other properties of wool.

During the first half of the previous century, staple crimp together with handle were used in appraising wool fineness [854], also referred to as wool quality or quality number, there being a broad inverse relationship between staple crimp and fibre diameter. Nevertheless, the relationship between staple crimp and fibre diameter (fineness) was not sufficiently reliable, and this practice has now largely fallen into disuse.

As single fibre crimp is difficult and time-consuming to measure in practice, staple crimp, bulk or resistance to compression, and more recently fibre curvature, are commonly measured and used as a measure of fibre crimp. It should be noted that fibre crimp, and therefore curvature, decreases with processing [855,856], in the top typically being 70–75% of that in the greasy wool [692,857], but it can be restored by steaming or wet relaxation, provided the sliver, top or roving has not been dyed [856]. Bulk resistance to compression can be measured after the wool has been scoured or converted into top, and provides a good measure of fibre crimp, provided an appropriate 'pre-relaxation' procedure is applied [858] and the effect of fibre diameter, or more specifically fibre stiffness, is allowed. Similarly, fibre curvature can be measured at any stage, by the OFDA or the Laserscan methods, provided appropriate fibre preparation and pre-relaxation, as necessary, procedures are applied.

In general, a measure of the crimpiness (crimp) in wool fibres can be obtained in one or more of the following ways:

- Single fibre crimp: On single fibres, by measuring the crimped (unstretched) and uncrimped (stretched) fibre lengths either by hand using a tweezer and ruler or by means of a sensitive tensile tester to determine when the fibre crimp has been removed, and then calculating the ratio of the uncrimped (stretched) length to the crimped (unstretched) length. Menkart and Detenbeck [456], for example, found a very good correlation between the staple crimp frequency in Merino wool and the stress required to uncrimp single fibres from the staple (i.e. unstretched).
- Staple crimp: On staples, either following a similar procedure to the above, or more commonly, counting, either by eye (i.e. manually) or image analysis (e.g. CSIRO Staplescan), the number of crimps per unit length (crimp frequency) present in the relaxed staple.
- Single fibre curvature: In this case, the fibre curvature is measured by image analysis or laser, for example, using instruments such as the OFDA100 or Sirolan-Laserscan™.
- Bulk/resistance to compression: The crimpiness of wool fibres is also popularly and reasonably estimated by measuring what is variously referred to as bulk, bulkiness, resistance to compression, bulk resistance to compression or compressibility of a randomised loose wool sample (bulk sample). Tests and associated instruments for measuring the bulk and resistance to compression of a randomised loose wool assembly have been developed in Australia [859], New Zealand [860] and South Africa [646]. Generally, the tests involve subjecting a fixed mass (weight) of clean wool fibres to an opening/randomisation and relaxation procedure, after which they are placed in a cylindrical container, and pre-compressed and their volume under a fixed pressure (or pressure at a fixed volume) measured. Bulk, or resistance to compression, can be described as the volume of a fixed mass of wool within a chamber (usually cylindrical), usually under a pre-determined pressure (or alternately, the pressure at a fixed volume) after a certain preparation (pre-relaxation) and pre-compression. The values so obtained, however, generally provide a measure of the product of fibre diameter and fibre (or staple) crimp.
- NIR technology: Wool bulk has also been estimated for NZ wool using NIR prediction equations [176,861].

Some of these methods have been described in more detail below.

2.7.2. *Staple crimp*

Staple crimp, for example, can be measured on a representative sample of unstretched greasy staples in the following manner:

- A 25 mm crimp glass is placed over the greasy staple and the number of crimps (or waves) per 25 mm counted and converted into crimps per centimetre.
- An image analysis-based system, such as the CSIRO Sirolan-Staplescan to measure staple crimp frequency [120] or the OFDA2000 is used.

Crowe [862] reported highly repeatable staple crimp frequency results obtained on three different automatic crimp meters, described previously [863,864], and which correlated well with manual measurements [865].

Figure 28. WRONZ auto wool bulkometer (www.newford.cn), reprinted with permission of IWTO.

2.7.3. Resistance to compression

The resistance to compression of a randomised fibre sample provides a measure of the bulk (volume) of wool lot under certain pressure. More than 60 years ago Van Wyk [866,867] laid the foundation for understanding the behaviour of a randomised bulk wool fibre sample under a compressive load, and the relationship between pressure and volume, as well as the influence of fibre properties and fibre arrangement within the bulk sample under compression. He showed that the bulk resistance to compression was closely related to the product of staple crimp frequency and fibre diameter. Various workers [868] have worked on the foundation that he laid.

Wool fibre crimp and curvature [856] decrease from the scoured fibre to top, and generally resistance to compression or bulk needs to be determined after some form of pre-treatment (e.g. steaming) to relax the fibres, this being particularly important once the fibres have been processed into sliver, tops or roving form, as such processing steps to a greater or lesser extent remove the fibre crimp.

Various instruments and test methods have been developed to measure the bulk resistance to compression of randomised wool samples, including the following:

- WRONZ loose wool bulkometer (Figure 28) and Auto-Bulkometer (Test Method in NZ-8716) [859,870–877]. WRONZ developed the manual loose wool bulk test, superseded by an automatic version, the core bulk test [860], which became a standard in New Zealand in 1994 [176]. Essentially, the bulk (volume) of a specific mass of wool is measured when compressed by a certain lead (or pressure).
- CSIRO compressibility tester (Figure 29) [859,878–882] measures the resistance to compression when a specific mass of wool is compressed to a specific volume.

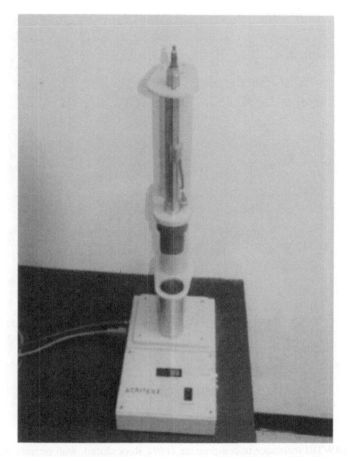

Figure 29. CSIRO compressibility tester (http://www.awta.com.au/en/Home/Publications/Newsletters), reprinted with permission of CSIRO.

- SAWTRI resistance to compression (Figure 30) [646,869,883], where essentially the height (in effect volume) of a specified mass of wool is measured when compressed under certain load (or pressure) as is the case of the WRONZ bulk test.

Swan and Mahar [859] have found the Australian and New Zealand measures of resistance to compression and bulk, respectively, to be related linearly, with both being extremely sensitive to sample preparation prior to measurement. Naik et al. [884] found a high correlation between the results of the WRONZ Bulkometer, tribocity and compression.

2.7.4. Curvature

Because of the documented importance of fibre crimp, preferably as part of objective measurement, attempts were made to find rapid cost-effective methods of measuring crimp in practice, these efforts leading to techniques involving instruments such as the OFDA and Laserscan for measuring single fibre crimp, actually curvature, simultaneously with diameter.

Figure 30. SAWTRI resistance to compression [869]. Reproduced, with permission, from J.P. van der Merwe, *The effect of fibre physical properties on woollen spun yarn and plain knitted fabric properties*, PhD thesis. University of Strathclyde, Glasgow, Scotland, 1985.

Edmonds [885] has reviewed the work done on the measurement of fibre curvature, while Swan and Mahar [886] discussed the metrology involved in the measurement of fibre crimp and curvature.

The first mention of fibre curvature measurement on the OFDA scheme was made in 1993 [341], while Dabbs et al. [887] reported for the first time to the IWTO in 1994 about the measurement of fibre curvature using the Laserscan, and in 1995 a report on the OFDA was presented to the IWTO.

In order to obtain a measure of fibre crimp, bulk etc., methods have been developed whereby the curvature of fibre snippets (generally 2 mm in length) is measured simultaneously with diameter, using the OFDA100 [341,356,856,888–894] and Sirolan-Laserscan™ [887–891,894] systems. Details of the two test methods and procedures and applications of results have been given in various publications, it being important that the fibres are pre-treated or 'relaxed' in a standard way, as this, and the fibre and snippet sampling and preparation techniques (e.g. scouring, mini-coring etc.), can affect the fibre curvature results to a greater or lesser extent [337,339,373,377,391,888,895–913], as also can the wear of the mini-coring tubes and the repeated mini-coring [895,897]. Scouring of top samples was found to increase fibre curvature more than immersion in hot water (with rubber bands holding the ends of the top sample together), the trends being similar for the OFDA and Sirolan-Laserscan™ [377,901]. Scouring increased curvature by about 12°/mm for low curvature (30–40°/mm) wools and about 25°/mm for high curvature (60–100°/mm) wools [377], part of the effect possibly being due to the fact that the scoured tops were mini-cored (as opposed to guillotined) producing shorter and more varied snippet lengths [377,901–906].

Various studies have been undertaken on the relationship between fibre curvature and other measures of fibre and staple crimp, as well as other factors, notably fibre snippet length, preparation and relaxation, which can influence the fibre curvature measurement. Swan [899,911] had concluded that staple crimp frequency is primarily a measure of the single fibre curvature. It has been found that the results correlated well with staple fibre crimp (90%) and wool bulk (85%) and resistance to compression (92%). Greasy fibre curvature was found to be correlated ($R^2 = 0.80$) with greasy staple crimp frequency [120,897], 1 crimp/cm corresponding approximately to 10°/mm in curvature. Smuts et al. [912] investigated the role of sheep breed on the OFDA100 curvature versus staple crimp frequency relationship. Van Rensburg [894] observed differences between the OFDA and Sirolan-Laserscan™ measurements of curvature for South African wools.

Fish [896] and Swan [911] concluded that there were potential limitations in the application of fibre curvature measurement for the specification of raw wool, including the relative instability of the measurement, the need for standardisation and industry acceptance. The issue of instability could be addressed through calibration, standardisation, relaxation and understanding the process involved in measurement and the factors that influence the results, such as processing and reducing crimp/curvature. Ways of improving the accuracy and consistency of curvature measurements on the OFDA100 and Sirolan-Laserscan™ systems by using calibration tops were reported by Fish (http://www.awta.com.au). Marler [905] reported the steps to harmonise the fibre curvature measurements on the OFDA and Sirolan-Laserscan™ systems, recommending [905,909] that Interwoollabs develop mean fibre curvature values for its fibre diameter standards, as reference values, and that IWTO-47 be modified to include calibration for mean fibre curvature and the reporting of results.

Curtis and Stanton [834], using the Sirolan-Laserscan™ method, found the curvature values for Australian sale lots to vary from 58°/mm to 167°/mm, with a mean value of 94.4°/mm.

Ponzoni et al. [913] gave the following grading for fibre curvature:

- Low curvature: ≤50°/mm.
- Medium curvature: 60 to 90°/mm.
- High curvature: >100°/mm.

Swan and Mahar [914] outlined a method for measuring fibre crimp/curvature based on light microscopy and image analysis, also briefly reviewing the measurement and importance of fibre crimp. They found that fibre curvature distribution is skewed, but simple transformations can provide approximately normal distributions, leading to highly repeatable and precise results, allowing curvature to be measured throughout the processing pipeline. Various papers [860,897] have reviewed aspects relating to the definition, measurement and importance of fibre curvature and crimp. Greeff and Schlink [915], for example, showed that felting (felt ball density) increased as fibre curvature and diameter decreased (and yield increased). Crook et al. [916] found that greasy wool curvature variation did not provide a useful indirect measure of visually assessed crimp definition.

Table 14. Standard thickness (diameter) limits and number of crimps per inch [927]. Reprinted from J.E. Deurden, J. Text. Inst. 20 (1929) pp. T93–T100, with permission of The Textile Institute.

Quality number	Thickness limits (μm)	Crimps per inch	Crimps per cm
Above 120s	14.0–14.7	28–30; 27–30	11–12
120s	14.7–15.4	25–27; 24–27	9–11
100s	15.4–16.2	22–24; 21–24	8–9
90s	16.2–17.0	20–21; 19–21	7–8
80s	17.0–17.9	18–19; 17–19	7
70s	17.9–18.9	16–17; 15–17	6–7
66s	18.9–20.0	14–15; 13–15	5–6
64s	20.0–21.3	12–13; 11–13	4–5
60s	21.3–23.0	10–11; 9–11	4
58s	23.0–25.5	8–9; 7–9	3–4
56s	25.5–29.0	6–7; 5–7	2–3

2.7.5. Interrelationship between fibre crimp and other fibre properties

2.7.5.1. Interrelationship between staple crimp, diameter and resistance to compression.
In pioneering work on the bulk resistance to compression of loose wool, Van Wyk [866,867] showed that the bulk resistance to compression of a randomised wool sample was related to the product of the staple crimp frequency and fibre diameter. Several other workers have subsequently confirmed and extended his general findings [38,454,646,883,917–925]. Wilkins et al. [926] found that for superfine Australian wool the resistance to compression was influenced by crimp frequency, and to a lesser extent by diameter.

Deurden [927], working on South African Merino wool, was one of the first workers to establish the broad relationship between staple crimp and fibre diameter (Table 14), this subsequently being confirmed by Lang [928,929] on Australian wool (Table 15). The following table (Table 14) gives the numerical relationships that Deurden found between the quality number, thickness limits and staple crimp.

Lang [928] and Whiteley and Charlton [930] also demonstrated the relationship between staple crimp and diameter for Australian wool. The following table (Table 15) being derived from the results given by Lang [928].

Within a mob of fine-wool sheep, finer fleeces do not necessarily have higher staple crimp [931], whereas for sale lots and consignments there is the well-known broad inverse relationship between mean fibre diameter and staple crimp frequency [575].

Table 15. Mean fibre diameter, staple crimp and quality number for Australian wool [928]. Reprinted from W.R. Lang, J. Text. Inst. 38 (1947) pp. T257–T270, with permission of The Textile Institute.

	Quality number						
	56s	58s	60s	64s	66s	70s	80s
Staple crimp/inch	6.1	7.9	9.8	12.5	14.1	16.4	19.2
Mean fibre diameter (μm)	26.5	23.5	20.9	19.7	19.7	18.9	15.6

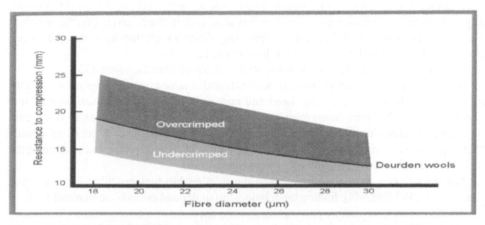

Figure 31. An example of a reference curve for the resistance to compression of steamed tops, or laboratory-scoured raw wool, based upon wool obeying Deurden's diameter–crimp relationship (SAWTRI Test) [116,934]. Reproduced with permission, from L. Hunter, *The effects of wool fibre properties on processing performance and yarn and fabric properties*, in *Proceedings of the 6th Int. Wool Text. Res. Conference*, 1, Pretoria, 1980, pp. 133–193 and L. Hunter, *A summary of SAWTRI's research on wool and wool blends – 1952–1987*, SAWTRI Spec. Publ. WOL 78, Port Elizabeth, South Africa, November 1987.

Smuts et al. [932] found that the bulk/diameter ratio (resistance to compression divided by diameter) had a good correlation with staple crimp and defined this as a measure of crimpiness. Whiteley et al. [933] confirmed the strong correlation between resistance to compression and crimp characteristics [923].

Hunter [116,934] showed (Figure 31) how one could distinguish between over-crimped and under-crimped wools (relative to Deurden values) on the basis of their resistance to compression and diameter values, this being important when acquiring tops for which staple crimp values are not available.

In view of the high correlation between resistance to compression of bulk wool samples and the crimp characteristics of the fibre, it can be reasonably assumed that many of the results and trends that apply to the one, probably will apply equally to the other [116], provided, as already mentioned, that allowance is made for the effect of fibre diameter (stiffness) on bulk resistance to compression. It has been suggested [116,447,562,883,921, 933] that, because of the good relationship between resistance to compression and certain processing characteristics, the former should be introduced into the objective measurement programme for wool.

2.7.5.2. Interrelationship between single fibre crimp, curvature, resistance to compression (bulk) and diameter. Smuts et al. [932], using the data from 18 tops, varying in crimp and diameter, found that within the top samples single fibre crimp was seldom significantly related to fibre diameter. In a few instances where significant trends were found, the correlation coefficient was relatively low. In all such significant cases, however, crimp tended to increase with an increase in fibre diameter within a sample, which contrasts with the between-sample trend, and agrees with the findings of Dusenbury and Wakelin [935] and Collins and Chaikin [689].

Whiteley and Charlton [936] found that there was a highly negative correlation between single fibre crimp frequency and mean fibre diameter, both between and within the samples. O'Connel and Lundgren [937] found that the crimp/diameter relationship differed between sheep of the same breed, and between different breeds.

Longer staple lengths tend to be associated with lower fibre curvature [23].

Sumner and co-workers [938–943] demonstrated a strong relationship between fibre curvature and diameter on the one hand and core bulk measured by means of the New Zealand Auto-Bulkometer on the other hand. Sumner and Upsdell [938] found a significant curvilinear relationship between fibre curvature and core bulk over a series of breeds differing markedly in curvature, and a linear relationship between fibre diameter and core bulk. Within a breed, both relationships were linear [938].

Various workers have also investigated the relationship between fibre curvature and diameter [23,897,898,944], finding the relationship to depend on wool source and type.

Within a fleece, mean fibre curvature can either decrease or increase with an increase in mean fibre diameter, depending upon the site of the sample [897,936,938]. McGregor and Toland found that for a commercial low staple crimp frequency fine-wool merino flock fibre curvature decreased by 2°–3°/mm for each 1 μm increase in mean fibre diameter. Reduced fibre curvature was also associated with increased staple length, the latter increasing by 3.3 mm for each 1 μm increase in fibre diameter and decreasing by 2.8 mm for each 10°/mm increase in fibre curvature. They surmised that the observed increases in the staple length/mean fibre diameter ratio with decreasing fibre curvature could be due to differences in the level of helical fibre crimp, because helical crimped fibres are longer than sine-crimped fibres. Other papers [945,946] deal with aspects relating to crimp.

In addition to crimp frequency, crimp form has also received attention. Chaudri and Whiteley found that single fibre crimp is often a combination of sine and helical forms and fibres from the same staple can vary greatly in crimp form [947]. Hunter [116] was of opinion that, in practice, differences in wool fibre crimp form do not significantly affect resistance to compression. The role of crimp amplitude, as opposed to crimp frequency, in curvature and processing behaviour still needs to be investigated.

2.7.5.3. Relationship between crimp and strength. Smuts et al. [932] concluded that for individual results obtained within top samples, single fibre tenacity and extension at break were, for all practical purposes, constant and independent of crimp and fibre linear density. For the sample means (i.e. between samples), however, an increase in crimp was associated with a decrease in fibre tenacity. In another study [646] of 200 South African tops, it was again found that between tops bundle tenacity decreased with an increase in staple crimp frequency. Huson [948] demonstrated the adverse effect of set and curvature on Lincoln wool fibre tensile properties, also briefly reviewing previous work on the effect of natural and artificial crimp on wool fibre tensile properties, the adverse effect of crimp being related to both the physical as well as the chemical nature and structure of the crimped fibre.

2.8. Style

Rogan [9] stated that style is a composite characteristic of greasy wool, usually being subjectively characterised in terms of the following: staple tip weathering, dust penetration into the staple, greasy wool yellowness, staple tip shape and crimp definition, the first mentioned two characteristics varying most between style grades [949]. Many of these factors, individually and collectively, affect processing performance. For example, most, if not all, of the weathered staple tip is removed during carding and combing [950,951]. Although the results are not always consistent, wools of better style tend to process better than those of poorer style [541,575,624,899,911,952–956].

As early as in 1964, Paynter [957] discussed the importance of wool style and physical fleece characteristics in determining wool price, while Jackson and Rottenbury discussed the measurement of style (style metrology), Winston [958] also discussed the objective measurement and processing consequences of wool style and type.

Wool style has been traditionally assessed subjectively, involving elements of crimp frequency (higher values assessed as being better), crimp definition and range, dust penetration, tip length and colour [692], many of these elements not surviving scouring.

For Australian Merino fleece wools, there are seven style grades [9,949,959], namely, choice, superior, spinners, best top-making, good top-making, average top-making and inferior top-making grades, these normally being subjectively assessed. Style depends on both genetics/breeding (in terms of crimp frequency, character, yolk composition and staple structure) and the environment (e.g. dust, water staining, VM, temperature effects on yolk colour, weathering, UV damage etc.) [960]. Style can be improved by covering sheep with coats, which reduces both dust levels and weathering [960–962], and beneficially affects wool quality and processing [623,763,963].

Following the development of the various objective and additional measurement tests and test methods, the instrument measurement of style, or at least the components of style (visual and tactile attributes), was virtually all that remained for the full characterisation of greasy wool in terms of its textile quality and processing behaviour and performance.

The first attempts to apply image analysis to objective measurement of the visual characteristics, notably crimp frequency and definition of greasy wool staples, were reported by Higgerson and Whiteley [964,965]. Jackson and Rottenbury [949] discussed the metrology, which underpins style measurement, based primarily on staple imaging technology. They gave a detailed description of the factors related to style and its objective (instrument) and subjective assessment. The measurement of style needs to provide information on the nature of the staple tip (e.g. depth or length, blocky or tippy etc.), crimp frequency and definition, basically the alignment of fibres in the staple, clarity and uniformity of the crimp along and across the staple (cotted wools represent an extreme case of poor crimp definition, being tender wool in which there has been an entanglement of loose fibre ends or tenuous fibres) and the presence and extent of dust, VM, weathering and coloured zones in the staple [949]. No relationship has been observed between fibre diameter variability and 'staple character' or crimp definition [966].

The following style-related measurement traits have been listed [911,949,954,967]:

- Staple length: The length of a staple in millimetres, projected along its axis without any stretch or crimp removal.
- Tip length (degree to which the staple has a weathered or washed-out zone at the staple tip): The distance in millimetres from the staple tip to the point where the staple width is 75% of the average width.
- Staple crimp frequency: Number of crimp waves per centimetre of staple length.
- Crimp definition (crimp amplitude and fibre alignment).
- Wool yellowness, greasy yolk colour and whether there are stain bands.
- Wool area.
- Dust area (dust penetration into the staple).

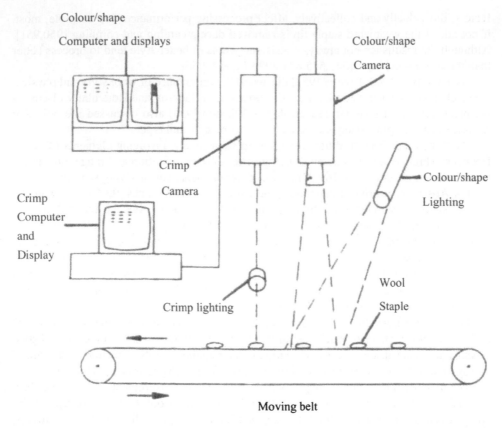

Figure 32. Schematic diagram of the laboratory prototype of the style measurement system [968]. Reprinted from R.D. Wright, D.W. Crowe and R.A. Rottenburg, *The objective measurement of style and type*, in *Proceedings of the 8th Int. Wool Text. Res. Conference*, II, Christchurch, New Zealand, 1990, pp. 58–67, with permission of CSIRO.

- Dust colour.
- The presence and type of VM.

The colour of wool staples in style measurement refers to the average colour (CIE*X; Y; Z tri-stimulus values and yellowness (Y–Z)) of the whole staple.

In the 1990s the CSIRO developed instruments [954,968] to measure wool style and type. The CSIRO style instruments (Figure 32) were developed to quantify the remaining visually assessed qualities [85], including greasy colour, crimp frequency, crimp definition, tip (shape and weathering) and dust penetration [969], although only tip shape plays a minor role in wool from adult sheep [624]. The style instruments provided quantitative estimates [970] of staple length, tip length, crimp frequency and definition, greasy wool yellowness, wool area, dust area and dust colour, all of which are components in the subjective assessment of style. Hansford [971] evaluated the performance of the style instruments. Hansford et al. [972] demonstrated the feasibility of the reliable objective measurement of the style components (staple length, tip length, crimp frequency and staple colour) of greasy wool using the automated image analysis instruments developed by the CSIRO.

Humphries [973] detailed the work carried out at the CSIRO to develop an instrument for the objective measurement of style, and gave a detailed description of the instrument, which essentially comprised a line-scan camera for measuring crimp details and a colour video camera for measuring colour and shape, together with the necessary computer hardware and software algorithms to analyse the data generated by the two cameras. The instrument used image analysis to produce a visual description of greasy wool staples, which are conveyed to a video camera on a conveyer belt. The CSIRO style instruments can predict the style grade which a skilled appraiser would assign a particular wool lot [949,963,974], providing an 'objective-style grade'. A final report on the performance of the style instruments was presented [967,970] in November 2000, which concluded that the instruments measure characteristics that could discriminate between commercial sale lots but more work was necessary in terms of robustness for commercial use. No link between diameter variability (SD) and style factors was observed [179]. The AWTA Ltd and Woolmark Company in 2002 presented a final report on the performance of the style instruments [970], basically concluding that the instruments were not commercially suitable for measurements in their current form. Each system gave different results due to its lighting system, although the data could be used to discriminate between sale lots, further work being suspended indefinitely.

Stevens and Crowe [624] concluded that it is inappropriate to use the percentage dust in staples as an indicator of fibre damage or processing wastes.

Cloete and De Beer [975] reported a positive association between mean fibre diameter and CV_D on the one hand and subjectively assessed quality, regularity of crimp and colour on the other hand, for South African commercial wool sheep flocks, confirming results obtained previously on research flocks [976,977].

2.9. Miscellaneous tests

In addition to the tests already discussed, various other tests have been developed for measuring wool characteristics that have not been measured, or effectively measured by the other tests. Most of these tests have never progressed beyond the laboratory stage or else are used only in laboratories. Some of these are briefly discussed below.

Because of the importance attached to the entanglement of scoured wool in terms of fibre breakage during carding, various methods for measuring entanglement in scoured wool and slivers have been developed and evaluated [32,974,978–982] in terms of their relationship to subsequent processing behaviour. Specific opening energy was shown to be related linearly to noil and inversely to Hauteur [983], and more entanglement during scouring leading to a lower Hauteur and more noil, thereby indicating more fibre breakage during processing. Harrowfield [984] stated that top Hauteur could be increased by approximately 10 mm and noil reduced by 1 to 2% absolute if fibre entanglement prior to carding could be avoided.

The Fibre Profile Meter [985] was developed to measure the fibre length and fibre diameter characteristics, including diameter profile, along individual fibres. Following the previous works [419,424,427,986–990] on fabric prickliness, He and Wang, [988] studied the effect of within fibre diameter variation on the flexural buckling behaviour, and consequently on prickle, showing that, as could be expected, the critical buckling load of the fibres decreases with increasing along fibre diameter variation.

Peterson et al. [889] described the application of the Single Fibre Analyser (SIFAN) for measuring the diameter profile of single wool fibres. Wang et al. [287,390] reported on the measurement-precision and evaluation of the diameter profiles of single wool fibres, using the SIFAN 3001, concluding that measuring randomly sampled 50 fibres from a mid-side sample is sufficient to provide an average diameter profile of fibres within a staple. The precision for the mean fibre diameter was of the order of 1.3 μm. The SIFAN enables the fast and simultaneous measurement of the diameter profile and the breaking force of single fibres [889,991], while the Sirolan-Tensor™ [992] measures fibre bundle strength by using a 3 mm gauge length.

Schaare and Jordan [95] have discussed the application of various measurement technologies, including fibre measurement technologies, and their impact on animal production.

Various attempts have been made to develop alternative (unconventional) technologies for estimating the wool yield of greasy wool [993–995] using, for example, nuclear-based techniques, such as X-rays and neutrons. Baxter and Wallace [993] reported on an innovative X-ray-based portable yield instrument for measuring greasy wool samples, the further development of which had been terminated because of various factors, including escalating cost.

Abbott and co-workers [996] discussed the different types of contamination encountered in wool, and ways of dealing with them, including their sources, detection and removal. Iype and Wan [997] developed an image processing method to analyse the structure and identify neps in slivers produced on the worsted system, briefly also reviewing the work done on neps, nep being a term used to describe a tangled 'knot-like' mass of unorganised fibres, counted and classified according to the size of the core in millimetre. The testing and processing consequences of neps in worsted processing have been discussed by Robinson et al. [998], nep levels decreasing during processing into yarn, being almost undetected after twisting and fabric finishing, dyeing (particularly dark blue) and finish masking such faults. Finer fibres are more inclined to entanglement and nep formation than coarser fibres. Vegetable matter and nep testing on tops (or slivers) is carried out automatically on the optalyser. Harrowfield [984] stated that neps in tops do not affect spinning performance and yarn properties in a direct way, but are undesirable in wool products.

Ranford et al. [999] discussed the application of visible/near infrared analysis in the textile industry, while Wood [176] discussed the application of NIR for the measurement of various greasy and scoured wool characteristics, such as moisture content, residual grease level, bulk, medullation, ethanol extraction (as an alternative to soxhlet ethanol extraction) and, together with the visible part of the spectrum, 'as –is' colour. Keogh and Roberts [258] had some success in the application of NIR to measure the yield and mean fibre diameter of mid-side samples. Other papers [1000] have also reported on the application of NIR for the analysis of wool. Hammersley et al. [1001] discussed problems with the NIR analysis of wool associated with calibration transfer.

3. Concluding remarks and further reading

This review undoubtedly demonstrates the enormous amount of research done over decades that aimed at developing test methods to objectively, accurately and cost-effectively measure those raw wool characteristics which determine the price, textile processing performance and product quality and to implement them in objective measurement programmes. Great progress has been made in this regard to standardise the test methods and place them on a sound scientific and cost-effective basis. Nevertheless, research continues with a view to improve existing test methods and develop new ones, some of which are aimed at measuring other raw wool characteristics, such as style, which play a role in the price, textile performance and product quality of wool. In the past, maximum attention has been devoted to the accurate and cost-effective measurement of fibre diameter and its distribution, with laser and image analysis-based methods, particularly the former being the preferred one today. The measurement of staple strength and length and raw wool (scoured) colour has also been standardised and fairly widely implemented and accepted in objective measurement programmes.

Although the measurement of wool base, VM base and VM type has, over many years, been improved, streamlined and standardised and is accurate, the underlying technologies used are dated and are of 'old generation'; there is definitely a need to develop more advanced, preferably non-destructive and eco-friendly, twenty-first century test methods. In this respect, some pioneering work based upon using different parts of the electromagnetic spectrum, such as X-rays and NIR, has been undertaken, but has as yet not achieved the desired accuracy. Image analysis and laser-based measurements of single fibre curvature, simultaneously with diameter, appear to provide an acceptable and accurate measure of fibre crimp, although careful attention has to be given to sample preparation, because fibre crimp is easily changed by external factors, such as tension, moisture and heat, which necessitates standard relaxation, specimen preparation and testing procedures for obtaining accurate and reliable measure of crimp. Nevertheless, further work is required before it is universally accepted as a routine part of objective measurement. Although image analysis-based methods applied to wool staples, for example during the staple length and strength test, have shown considerable potential for providing an acceptable measure of style, this test has, as yet, not received widespread acceptance, and further development work appears to have stopped.

Due to their low levels, i.e. the proverbial 'needle in the haystack', practical, accurate and cost-effective test methods for measuring dark (coloured) and medullated fibre (notably objectionable medullated fibre) contamination, as part of the objective measurement pro-gramme for raw wool, appear to be very difficult, if not impossible, to achieve the required standards within the near future.

Over the past two decades considerable progress has been made with the 'on-farm' measurement of certain raw wool characteristics, notably mean fibre diameter and diam-eter profile, using highly advanced technologies. Such tests are, however, not intended or proposed for certification purposes. Nevertheless, further work in this field is continuing, and it is not impossible that the future could see accurate and cost-effective 'on-farm' test methods being developed for measuring most wool characteristics which are presently measured in 'central' objective measurement laboratories. The ultimate goal would be that each farmer could accurately measure and 'certify' the important properties of the wool in each bale of wool being offered for sale. It is not too far fetched to see this being achieved within the next 20–30 years. Essentially, this would require staple, representative of the fleeces forming the contents of a bale of wool, to be 'scanned' to provide accurate and reliable measures for the following properties:

- Staple length and length variability.
- Fibre diameter, including diameter profile and distribution, which could also provide a measure of staple (or rather fibre) strength.
- Wool content/yield.
- Staple crimp and crimp definition, as well as fibre alignment.
- Wool colour.
- Staple tip length (depth) and colour.
- VM content and type.

The above may involve some form of staple cleaning procedure and/or employing dif-ferent parts of electromagnetic spectrum, as well as highly sophisticated farm-management practices, where each sheep has an identity and wool characteristic (quality) profile, possibly only comprehensively determined once and then software adjusted according to farming conditions, shearing date etc. Each bale would be able to identify the sheep from which the wool has come. This may become practically feasible as new technologies are developed and become cost-effective. In addition, some form of dark (coloured) and medullated (preferably objectionable medullated) fibre risk scheme assessment would be required at farm level.

The reader is referred to the following references for further reading:

General
[18], [85], [116], [117], [128], [129], [167], [216], [575], [1002], [1003]
Objective measurement and specification
[131], [132], [134], [164], [165], [168], [171], [175], [176], [536], [1003–1006],
Diameter measurement
[249], [262], [280], [283], [287], [388], [1007], [1008], [1009]
Colour and yellowing
[746], [774], [777], [780], [781], [791], [1010]
Dark and coloured fibres
[807], [810], [825]
Vegetable matter
[1011]
Weathering and style
[764], [972]
Fibre and staple length
[200], [607]
Fibre and staple strength
[226], [658], [1011], [1012]
Fibre bulk, crimp and curvature
[859], [860], [886], [890], [891], [895–897], [914], [933]
On-Farm measurement
[495], [503], [504], [999],
Clip preparation
[77], [78]

Acknowledgements

The authors would like to express their appreciation to Edna Hunter for assistance with the word processing and Jenny Wooldridge for assistance with the references.

References

[1] Anonymous, *Staple measurements*, AWTA Ltd. Newsletter, Victoria, Australia, October 2007.
[2] K.J. Whiteley, Wool Technol. Sheep Breed. 36(2) (1987) pp. 109–113.
[3] R.C. Couchman, P.J. Hanson, K.J. Stott and C. Vlastuin, *Wool Quality: implications for worsted processing, grower receipts and R&D*, Workshop on Management for Wool Quality in Mediterranean Environments, Western Australian Department of Agriculture, Perth, Australia, 1992.
[4] P. Graham, P. Abraham and J. Crean, Wool Technol. Sheep Breed. 43(2) (1995) pp. 111–123.
[5] R.C. Couchman, Wool Technol. Sheep Breed. 43(2) (1995) pp. 124–140.

[6] P.J.M. Bell, *Implications to top makers and wool combers of alternative clip preparation techniques*, in *Proceedings of the Seminar on Clip Preparation Research*, CSIRO Division of Textile Physics and Australian Wool Corporation, Ryde, Australia, April 1983, pp. 127–158.

[7] D. Charlton and H.G. David, Wool Technol. Sheep Breed. 35(2) (1987) pp. 131–141.

[8] P.R. Lamb and S. Yang, *The effect of wool fibre properties on spinning performance and yarn properties*, in *Proceedings of the Wool Spec 94, Seminar 'Specification of Australian Wool and its Implication for Marketing and Processing'*, CSIRO Division of Wool Technology and International Wool Secretariat, Sydney, Australia, November 1994, pp. R1–R11.

[9] I.M. Rogan, Wool Technol. Sheep Breed. 43(4) (1995) pp. 295–306.

[10] P.R. Lamb and D. Yang, Wool Technol. Sheep Breed. 45(4) (1997) pp. 283–308.

[11] C. Nolan, Wool Technol. Sheep Breed. 34(2) (1986) pp. 72–74.

[12] M.W. Andrews, *The objective measurement of raw wool in Australia*, in *Proceedings of the 11th Annual Conference, 'Measurement, Construction and Performance'*, Lincoln College, Canterbury, New Zealand, 25–27 November, 1983, pp. 41–57.

[13] R. Pattison, Wool Technol. Sheep Breed. 29(3) (1981) pp. 93–98.

[14] K.J. Whiteley, Wool Technol. Sheep Breed. 31(2) (1983) pp. 58–61, 64.

[15] P. Graham, Wool Technol. Sheep Breed. 51(2) (2003) pp. 130–133.

[16] G.D. Snowder, Wool Technol. Sheep Breed. 41(2) (1993) pp. 230–238.

[17] International Wool Secretariat, *Wool Premiums and Discounts Analysis of Wool Sold at Auction – Third Quarter 1994/95 Season*, International Wool Secretariat, Melbourne, Australia, 1995.

[18] P. Baxter, Wool Technol. Sheep Breed. 44(1) (1996) pp. 29–38.

[19] P.R. Lamb, K.M.S. Curtis, W. Humphries and N. Pant, Wool Technol. Sheep Breed. 50(4) (2002) pp. 679–684.

[20] A.A. Swan, L.R. Piper and I.W. Purvis, *Breeding options in fine wool flocks*, in *Proceedings of the Finewool 2000 Symp.*, CSIRO, Armidale, Australia, 2000, p. 65.

[21] P.R. Lamb and S. Yang, *The commercial impact of fibre properties in spinning*, IWTO Report No. 22, Dresden, Germany, June 1998.

[22] P.R. Lamb and S. Yang, *The advantages of longer Hauteur*, Report No. WT 96-01, CSIRO Division of Wool Technology, Geelong, Australia, 1996.

[23] P.R. Lamb and K.M.S. Curtis, *Identifying customer needs and addressing process and product opportunities*, Discussion paper, Australian Sheep Industry CRC, January 2004.

[24] C.J. Scrivener, A.L. Vizard and K.A. Hansford, Wool Technol. Sheep Breed. 47(4) (1999) pp. 241–247.

[25] T. Drummond, Wool Technol. Sheep Breed. 41(3) (1993) pp. 317–329.

[26] K.J. Whiteley, *The influence of fibre characteristics on processing and garment performance*, in *Proceedings of the 4th World Merino Conference*, Uruguayan Wool Secretariat, Uruguay, 1994, pp. 209–227.

[27] C.M. Oldham, R.C. Couchman and G.A. Robinson, *Report on an investigation into effects of season of shearing and selling centre on processing performance to top*, IWTO Report No. 6, Dresden, Germany, June 1998.

[28] A.M. Downes, Appl. Polymer Symp. 18 (1971) pp. 895–904.

[29] R.A. Rottenbury, D. Charlton and P.J.M. Bell, *Further processing studies on clip preparation: six clips of differing characteristics*, IWTO Report No. 8, Paris, January 1986.

[30] B.V. Harrowfield, *Worsted processing technology: An Overview*, in *Proceedings of the Woolspec 94, Seminar 'Specification of Australian Wool and its Implication for Marketing and Processing'*, CSIRO Division of Wool Technology and International Wool Secretariat, Sydney, Australia, November 1994, pp. P1–P8.

[31] J.M. Bennett, Wool Technol. Sheep Breed. 29(3) (1981) pp. 122–125.

[32] G.A. Robinson, *The nature of scoured wool and its preparation for carding*, in *Proceedings of the Wool Scouring and Worsted Carding: New Approaches, Symposium*, CSIRO, Geelong, Australia, November 1986, pp. 38–43.

[33] W.R. Lang, R. Henwood, Rankin, J.B. Wright and W.K. Campbell, J. Text. Inst. 58(5) (May 1967) pp. 185–193.

[34] P.J.M. Bell, D. Charlton and R.A. Rottenbury, Wool Technol. Sheep Breed. 34(4) (December 1986/January 1987) pp. 125–131.

[35] R.B. Whan, Wool Technol. Sheep Breed. 17(2) (1970) pp. 11–13.

[36] R.B. Whan, J. Text. Inst. 63 (1972) pp. 84–90.

[37] R.B. Whan, Wool Technol. Sheep Breed. 15(1) (July 1968) pp. 87–91.
[38] M.W. Andrews and R.A. Rottenbury, Wool Technol. Sheep Breed. 22(3) (1975) pp. 23–26.
[39] E.M. Pohle, D.D. Johnston, H.R. Keller and H.C. Reals, USDA Marketing Report No. 211, February 1958.
[40] G.R. Stanbury and W.G. Byerlely, J. Text. Inst. 25 (1934) pp. T295–T310.
[41] R.B. Whan, Wool Technol. Sheep Breed. 7(2) (1960) pp. 15–20.
[42] A. Mackenzie, Wool Technol. Sheep Breed. 27(1) (1979) pp. 11–14.
[43] B. Thompson, K.J. Whiteley, J.R. Black and D.C. Teasdale, Wool Technol. Sheep Breed. 31(1) (1983) p. 22.
[44] B. Thompson, K.J. Whiteley, J.R. Black and D.C. Teasdale, Wool Technol. Sheep Breed. 31(1) (1983) pp. 35–39.
[45] L.B. Johnston, Wool Technol. Sheep Breed. 33(3) (1985) pp. 84–85.
[46] R.E. Tucker, D.C. Teasdale and D.M. Knight, Wool Technol. Sheep Breed. 36(1) (1988) pp. 12–16.
[47] D. Charlton and H.G. David, Wool Technol. Sheep Breed. 41(2) (1993) pp. 185–198.
[48] J. Skillecorn, Wool Technol. Sheep Breed. 21(2) (1974) pp. 7–10.
[49] M.W. Andrews and R.A. Rottenbury, Wool Technol. Sheep Breed. 21(2) (1974) pp. 19–21.
[50] S.J. Welsman, Wool Technol. Sheep Breed. 23(3) (1976) pp. 14–16.
[51] P.H. Savage, Wool Technol. Sheep Breed. 26(1) (1978) pp. 13–15.
[52] H.W.M. Lunney, Wool Technol. Sheep Breed. 30(1) (1982) pp. 18–22.
[53] R.A. Rottenbury, D. Charlton and P.J.M. Bell, Wool Technol. Sheep Breed. 34(1) (1986) pp. 4–11.
[54] T.S. Fairhead, Wool Technol. Sheep Breed. 34(2) (1986) pp. 41–46.
[55] K.J. Whiteley, Wool Technol. Sheep Breed. 19(2) (1972) pp. 31–32.
[56] P.R. McMahon, Wool Technol. Sheep Breed. 35(1) (1987) pp. 30–32.
[57] R.B. Whan, Wool Technol. Sheep Breed. 35(1) (1987) pp. 33–37.
[58] F.K. Dalton, Wool Technol. Sheep Breed. 35(1) (1987) pp. 44–46.
[59] E.L. Jenkins, Wool Technol. Sheep Breed. 35(1) (1987) pp. 57–60.
[60] K.J. Whiteley, Wool Technol. Sheep Breed. 35(1) (1987) pp. 81–85.
[61] J. O'Sullivan, Wool Technol. Sheep Breed. 37(3) (1989) pp. 98–100.
[62] D. Charlton and H.G. David, Wool Technol. Sheep Breed. 39(4) (1991) pp. 129–135.
[63] D.J. Cottle, Wool Technol. Sheep Breed. 42(2) (1994) pp. 193–211.
[64] H.G. David and D. Charlton, Wool Technol. Sheep Breed. 36(1) (1988) pp. 4–11.
[65] K.J. Whiteley, Wool Technol. Sheep Breed. 35(1) (1987) pp. 51–52.
[66] K.G. Ponting, Wool Technol. Sheep Breed. 11(1) (1964) pp. 111–112.
[67] E.L. Jenkins, Wool Technol. Sheep Breed. 21(2) (1974) pp. 13–16.
[68] D. Charlton and M.G. McInerney, Wool Technol. Sheep Breed. 30(4) (1982) pp. 165–170.
[69] H.W.M. Lunney, Wool Technol. Sheep Breed. 33(1) (1985) pp. 35–37, 40.
[70] J. McKinnon, Wool Technol. Sheep Breed. 17(2) (1970) pp. 21–23.
[71] F.K. Dalton, Wool Technol. Sheep Breed. 19(1) (1972) pp. 65–67.
[72] P.R. McMahon, Wool Technol. Sheep Breed. 19(2) (1972) pp. 7–10.
[73] Anonymous, Wool Technol. Sheep Breed. 20(1) (1973) pp. 31–32.
[74] E.L. Jenkins, Wool Technol. Sheep Breed. 35(1) (1987) pp. 57–60.
[75] W. Stadler and R.I. Gillies, Wool Technol. Sheep Breed. 42(4) (1994) pp. 319–326.
[76] H.W.M. Lunney, Wool Technol. Sheep Breed. 35(1) (1987) pp. 64–68.
[77] R.A. Rottenbury, D. Charlton, L.J. Smith and J.R. Eley, Wool Technol. Sheep Breed. 35(3) (1987) pp. 156–165.
[78] Anonymous, *Proceedings of the Seminar 'Clip Preparation Research'*, CSIRO Division of Textile Physics & Australian Wool Corporation, Australia, (Editor: H.W.M. Lunney), April 1983, pp. 5–159.
[79] K.A. Hansford, Wool Technol. Sheep Breed. 50(4) (2002) pp. 168–180.
[80] K.A. Hansford, Wool Technol. Sheep Breed. 49(2) (2001) pp. 156–166.
[81] P.R. Byrne, T.J. Mahar and R.W. Conors, Wool Technol. Sheep Breed. 48(2) (2000) pp. 147–152.
[82] R.A. Rottenbury, D. Charlton, L.J. Smith and J.R. Eley, *Clip preparation and its influence on fleece characteristics and processing performance: A final report covering some varied clip types*, IWTO Report No. 4, Rio de Janeiro, Brazil, June 1987.

[83] K.J. Whiteley, *Fleece variability – Its impact on processing*, in *Proceedings of the Seminar on Clip Preparation Research*, CSIRO Division of Textile Physics and Australian Wool Corporation, Ryde, Australia, April 1983, pp. 27–44.
[84] P.R. Byrne, T.J. Mahar and R.W. Connors, Wool Technol. Sheep Breed. 48(2) (2000) pp. 147–152.
[85] P.R. Lamb, Wool Technol. Sheep Breed. 50(4) (2002) pp. 671–678.
[86] Anonymous, Wool Technol. Sheep Breed. 50(1) (2002) pp. 40–50.
[87] P.R. Byrne, T.J. Mahar and R.W. Connors, *Assessing the variability of classed fleece lines using a Uniformity Index*, IWTO Commercial Regulations and Contracts, Committee, Appendix 2, Nice, France, 1997.
[88] P.R. Byrne, T.J. Mahar and R.W. Connors, *Prediction of wool preparation category for classed fleece lines using a Uniformity Index*, IWTO Report No. CTF 05, Nice, France, 1998.
[89] AWTA, CSIRO and The Woolmark Company, Wool Technol. Sheep Breed. 50(1) (2002) pp. 40–51.
[90] K.A. Hansford, *The use of objective testing of bales in-store for lot matching. 1. Interlot wool matched for fibre diameter on the basis of Laserscan measurements*, IWTO Report No. CTF 06, Shanghai, China, May 2001.
[91] S.A.S. Douglas, (Reporting on work by Bob Couchman), Wool Rec. 162(3701) (2003) p. 14.
[92] R. Bazeley, M. Friend and A. Tunstall, Wool Technol. Sheep Breed. 51(1) (2003) pp. 49–62.
[93] J.-H. Dittrich, D. Klee and E. Finnimore, Int. Text. Bull. Yarn Forming 2/83 (1983) pp. 61–84.
[94] R.W. Ponzoni, Wool Technol. Sheep Breed. 28(2) (1980) pp. 19–22.
[95] P.N. Schaare and R.B. Jordan, Wool Technol. Sheep Breed. 38(1) (1990) pp. 32–39.
[96] I. Rogan, Wool Technol. Sheep Breed. 43(3) (1995) pp. 244–253.
[97] D.J. Cottle and P.J. Bowman, Wool Technol. Sheep Breed. 38(3/4) (1990/1991) pp. 114–118.
[98] R.A. Rottenbury, D.J. Allen and H.F.M. Van Schie, *Mill consignments and the impact of sale-lot characteristics*, IWTO Report No. 27, Istanbul, Turkey, May 1993.
[99] R.A. Rottenbury, D.J. Allen and H.M.F. Van Schie, Wool Technol. Sheep Breed. 41(1) (1993) pp. 199–215.
[100] D.E.A. Plate, *What are the Wool Characteristics that are of Importance to Wool Processors and End Users?* CSIRO, Division of Wool Technology, Geelong, Australia, 1990.
[101] J. Turk, Wool Technol. Sheep Breed. 41(1) (1993) pp. 51–55.
[102] T. Turk, Wool Technol. Sheep Breed. 50(4) (2002) pp. 622–625.
[103] G.W. Walls, Wool Technol. Sheep Breed. 21(1) (1974) pp. 27–31 and Obj. Meas. Symp. Australia, Paper No. 3 AWC Melbourne, 1974.
[104] M. Lipson, Wool Technol. Sheep Breed. 12(2) (1965) pp. 51–52.
[105] M. Lipson, Wool Technol. Sheep Breed. 19(2) (1972) pp. 11–15.
[106] A. Johnson, Wool Technol. Sheep Breed. 4(1) (1957) pp. 71–75.
[107] W. von Bergen, Wool Technol. Sheep Breed. 10(1) (1963) pp. 43–49.
[108] A. Johnson, Wool Technol. Sheep Breed. 2(1) (1955) pp. 101–102.
[109] N.F. Roberts, Wool Technol. Sheep Breed. 4(2) (1957) pp. 65–70.
[110] P.R. McMahon, Wool Technol. Sheep Breed. 5(1) (1958) pp. 21–24.
[111] S. Read, Wool Technol. Sheep Breed. 43(4) (1995) pp. 318–322.
[112] P.J.M. Bell, Wool Technol. Sheep Breed. 26(4) (1978) pp. 29–37.
[113] P.R. McMahon, Wool Technol. Sheep Breed. 9(2) (1962) pp. 17–24.
[114] P.C. Lee, *Quality Control Aspects of Worsted Processing*, IWS Technical Publication, Ilkley, United Kingdom, 1991.
[115] G.A. Carnaby and D.C. Maddever, Wool Technol. Sheep Breed. 33(2) (1985) pp. 56–63.
[116] L. Hunter, *The effects of wool fibre properties on processing performance and yarn and fabric properties*, in *Proceedings of the 6th Int. Wool Text. Res. Conference*, 1, Pretoria, 1980, pp. 133–193.
[117] D.J.J. Van Rensburg, *The Prediction of Wool Worsted Spinning Performance and Yarn Properties*, PhD Thesis, University of Port Elizabeth, South Africa, 1997.
[118] A.F. Botha, *The fibre diameter distribution, particularly the coarse edge, of South African wool, and its effect on textile performance*, PhD Thesis, Nelson Mandela Metropolitan University, Port Elizabeth, South Africa, 2005.
[119] D.J.J. van Rensburg and L. Hunter, *A review of the influence of certain raw wool characteristics on worsted processing*, Tex Report No. 4, CSIR Division of Textile Technology, Port Elizabeth, South Africa, March 1992.

[120] P.R. Lamb, *The impact of CV(D) and crimp or curvature on Hauteur and Romaine*, IWTO T&S Committee, IWTO Report No. CTF 02, Christchurch, New Zealand, May 2000a.

[121] G.A. Robinson, *Processing factors and product quality in early stage processing*, in *Proceedings of the Woolspec 94, Seminar 'Specification of Australian Wool and its Implication for Marketing and Processing*, CSIRO Division of Wool Technology and International Wool Secretariat, Sydney, Australia, November 1994, pp. Q1–Q13.

[122] D.S. Taylor, Pakistan Text. J. (1988), pp. 17–34.

[123] K.J. Whiteley, J. Text. Inst. 79 (1988), pp. 339–348.

[124] G.A. Robinson, Wool and Woollens India, (1989), pp. 41–47, also Seminar 'Research into Specification of Australian Wool', 28 April, 1989.

[125] K.J. Whiteley, G. Robinson and B. Harrowfield, *Profit through innovation in early stage processing*, in *Proceedings of the Top-Tech, '96' Symp.*, CSIRO Division of Wool Technology, Geelong, Australia, 1996, pp. 1–15.

[126] D.S. Taylor, Wool Technol. Sheep Breed. 33(4) (1985) pp. 130–146.

[127] P. Grosberg, *The mechanical processing of wool*, in *Proceedings of the 5th Int Wool Text Res Conference*, I, Aachen, Germany, September 1975, pp. 341–359.

[128] N.A.G. Johnson and I.M. Russell, (ed.), *Advances in Wool Technology*, Woodhead Publishing Ltd., Cambridge, England, 2009.

[129] W.S. Simpson and G.H. Crawshaw, (eds.), *Wool Science and Technology*, Woodhead Publishing Ltd., Cambridge, England, 2002.

[130] Objective Measurement Symposium (OMSA-74), Published by the Australian Wool Corporation, Melbourne, Australia, March 1974.

[131] B.P. Baxter, Wool Technol. Sheep Breed. 50(4) (2002) pp. 766–779.

[132] P.J. Sommerville, Wool Technol. Sheep Breed. 50(4) (2002) pp. 853–860.

[133] D.C. Teasdale, Wool Technol. Sheep Breed. 43(1) (1995) pp. 166–176.

[134] K.J. Whiteley and R.A. Rottenbury, *Research into specification and marketing of Australian greasy wool*, in *Proceedings of the 8th Int Wool Text Res Conference*, II, Christchurch, New Zealand, 1990, pp. 1–27.

[135] D.F. Booth, Wool Technol. Sheep Breed. 21 (1974) pp. 19–24.

[136] L.B. Johnston, Wool Technol. Sheep Breed. 40(4) (1992) pp. 117–130.

[137] S.A.S. Douglas, Wool Technol. Sheep Breed. 35(1) (1987) pp. 11–13.

[138] B.H. Mackay, Wool Technol. Sheep Breed. 35(1) (1987) pp. 14–22.

[139] B.H. Mackay, Wool Technol. Sheep Breed. 19(2) (1972) pp. 25–28.

[140] J. McK. Hamilton, Wool Technol. Sheep Breed. 20(1) (1973) pp. 61–66.

[141] B.H. Mackay, Wool Technol. Sheep Breed. 20(2) (1973) pp. 44–47.

[142] R.A. Rottenbury and M.W. Andrews, Wool Technol. Sheep Breed. 22(1) (1975) pp. 19–22.

[143] AWC Advisory Committee on Objective Measurement, Wool Technol. Sheep Breed. 28(3) (1980) pp. 11–17.

[144] D.J. Ward, Wool Technol. Sheep Breed. 35(1) (1987) pp. 23–29.

[145] W.S. Simpson, Wool Technol. Sheep Breed. 37(2) (1989) pp. 73–77.

[146] P.B. Wallace, Wool Technol. Sheep Breed. 40(3) (1992) pp. 83–84.

[147] B. Mitchell, Wool Technol. Sheep Breed. 51(2) (2003) pp. 150–158.

[148] K.J. Whiteley, Wool Technol. Sheep Breed. 22(1) (1975) pp. 5–7.

[149] D.J. Ward, Wool Technol. Sheep Breed. 25(4) (1977) pp. 23–25.

[150] D. Provost, Wool Technol. Sheep Breed. 31(2) (1983) pp. 56–57.

[151] R.J. Quirk, Wool Technol. Sheep Breed. 31(2) (1983) pp. 43–48.

[152] J. Skillecorn, Wool Technol. Sheep Breed. 31(2) (1983) pp. 49–54.

[153] K. Baird, Wool Technol. Sheep Breed. 32(2) (1984) pp. 51–54, 58.

[154] R.D. Pattison and K.J. Whiteley, Wool Technol. Sheep Breed. 32(4) (1984) pp. 181–189, 196.

[155] D. Teasdale, Wool Technol. Sheep Breed. 34(4) (1986) pp. 139–140.

[156] T.S. Fairhead, Wool Technol. Sheep Breed. 35(1) (1987) pp. 91–96.

[157] J. Hoadley, Wool Technol. Sheep Breed. 39(3) (1991) pp. 102–105.

[158] K. Hansford, *Raw wool specification*, in *Proceedings of the Top-Tech '96 Symp.*, Geelong, Australia, 1996, pp. 18–20.

[159] D.J. Ward, Wool Technol. Sheep Breed. 35(1) (1987) pp. 61–63.

[160] R.J. Quirk, Wool Technol. Sheep Breed. 35(1) (1987) pp. 69–74.

[161] J. Skillecorn, Wool Technol. Sheep Breed. 35(1) (1987) pp. 75–80.

[162] K. Baird, Wool Technol. Sheep Breed. 35(1) (1987) pp. 86–90.

[163] D.J. Ward, Wool Technol. Sheep Breed. 16(2) (1969) pp. 17–23.

[164] K.J. Whiteley and R.A. Rottenbury, Wool Technol. Sheep Breed. 38(3/4) (1990/1991) pp. 83–88.

[165] A.R. Edmunds, *A Review of New Zealand R&D on raw wool specification for marketing*, in *Proceedings of the 8th Int Wool Text Res Conference*, II, Christchurch, New Zealand, 1990, pp. 18–27.

[166] D. Teasdale, Wool Technol. Sheep Breed. 37(1) (1989) pp. 44–48.

[167] Australian Wool Testing Authority Ltd., *Testing the wool clip*, Available at www.awta.com.au, (A.B.N. 43 006 014 106), 2009.

[168] P.D. Morgan, J. Text. Inst. 87(Part 3) (1996) pp. 34–42.

[169] J.W. Marler, *Interwoollabs Report On 1998 Round Tests*, IWTO Report No. SG 02, Florence, Italy, May 1999.

[170] R.L. Thompson and M.W. Andrews, *Instrumentation for sale by description*, in *Proceedings of the Fourth Wool Conference*, Metlan '86, Part 1, Gdynia, Poland, 10–11 June, 1986, pp. 111–123.

[171] Anonymous, *Objective measurement of wool in Australia*, Final Report of the Australian Wool Board's Objective Measurement Policy Committee, York Press Ltd., Hoddle Street, Abbotsford, 3067, pp. 1–112.

[172] J.W. Marler, Wool Technol. Sheep Breed. 40(3) (1992) pp. 85–88.

[173] B.C. Couchman, Wool Technol. Sheep Breed. 42(1) (1994) pp. 41–50.

[174] T. Gleeson, M. Lubulwa and S. Beare, Wool Technol. Sheep Breed. 41(4) (1993) pp. 394–405.

[175] E.J. Wood, G.R. Burling-Claridge, S.L. Ranford, M.J. Hammersley, A.R. Edmunds and B.L. Thomas, *Recent developments in the objective measurement of New Zealand wool*, in *Proceedings of the 9th Int Wool Text Res Conference*, IV, Biella, Italy, 1995, pp. 44–52.

[176] E. Wood, Wool Technol. Sheep Breed. 47(3) (1999) pp. 218–229.

[177] T.J. Mahar, Wool Technol. Sheep Breed. 37(1) (1989) pp. 20–26.

[178] W. Li, W. Humphries and S. Yang, *Increasing use of objective measurement in the Chinese domestic wool industry*, IWTO Report No. CFT 02, Shanghai, China, May 2001.

[179] M.R. Bow and K.A. Hansford, *Fibre diameter variability*, in *Proceedings of the Woolspec 94, Seminar 'Specification of Australian Wool and its Implication for Marketing and Processing'*, CSIRO Division of Wool Technology and International Wool Secretariat, Sydney, Australia, November 1994, pp. L1–L25.

[180] B.H. Mackay and H.G. David, Wool Technol. Sheep Breed. 12(1) (1965) pp. 57–64.

[181] H.G. David, Wool Technol. Sheep Breed. 14(2) (1967) pp. 39–43.

[182] R.A. Rottenbury and M.W. Andrews, Wool Technol. Sheep Breed. 22(1) (1975) pp. 37–40.

[183] H.G. David, Wool Technol. Sheep Breed. 27(4) (1979) pp. 35–37.

[184] A.B. Baumann, A.J. Burke and D.J. Allen, Wool Technol. Sheep Breed. 39(2) (1991) pp. 55–60.

[185] B.C. Russel and D.J. Cottle, Wool Technol. Sheep Breed. 41(2) (1993) pp. 173–184.

[186] G.W. Snedecor, *Statistical Methods*, 5th ed., Lowa State University Press, Ames Iowa, USA, 1962.

[187] G.E. Noether, *Introduction Statistics. The Nonparametric way*, Springer-Verlag, New York, 1991.

[188] A.R. Gilmour and K.D. Atkins, Australian J. Agric. Res. 43 (1992) pp. 1777–1788.

[189] H.W.M. Lunney and G.H. Brown, Text. Res. J. 53 (1983) pp. 281–289.

[190] H.W.M. Lunney and G.H. Brown, Text. Res. J. 55 (11) (1985) pp. 671–676.

[191] P.R. Lamb, G.A. Robinson and T.J. Mahar, *The effect of fibre crimp on yarn evenness and spinning performance*, in *Proceedings of the Top-Tech, '96' Symp.*, CSIRO Division of Wool Technology, Geelong, Australia, 1996, pp. 324–331.

[192] R. Henon, *Statistical contribution to the methods of classifying the fineness of wool, (In French)*, IWTO Report No. 15, London, England, June 1952.

[193] F. Montford, Text. Res. J. 30 (1960) pp. 556–567.

[194] F. Montford, *Scientific Aspects of the Woollen Industry, (In French)*, Dunod, Paris, France, 1960.

[195] H. Linhart and A.W.G. Van Der Westhuyzen, J. Text. Inst. 54 (1963) pp. T123–T127.

[196] H.W.M. Lunney and G.H. Brown, Text. Res. J. 55 (1985) pp. 671–676.

[197] L. Wang and X. Wang, Text. Res. J. 68 (1998) pp. 87–93.

[198] A.M. Ford, *A study of fibre length and diameter statistics*, WRONZ Communication No. C112, October 1989.

[199] V.E. Fish, *A review of the distribution of fibre length in tops*, IWTO Report No. SG 01, Hobart, Australia, April 2005.

[200] O. Mayo, B. Crook, J. Lax, A. Swan and T.W. Hancock, Wool Technol. Sheep Breed. 42(3) (1994) pp. 231–236.

[201] M.R. Fleet, P.A. Irvine, R. Ponzoni and M.R. Bow, Wool Technol. Sheep Breed. 30(3) (1982) pp. 139–149.

[202] E. Gee, *The use of skewness and kurtosis in the prediction of processing performance and yarn and fabric properties*, in *Proceedings of the 7th Int Wool Text Res Conference*, Vol. II, Tokyo, Japan, 1985, pp. 85–94.

[203] L.D. Bemoth and J.W. Marler, *Prediction of DMC extractable matter by NIR technology*, IWTO Report No. STG 01, Nice, France, December 2002.

[204] M.J. Hammersley, S.L. Ranford and P.E. Townsend, *Calibration of Near Infred Analser using dicloromethane sixlet results*, IWTO Report, No. 18, Cape Town, South Africa, April 1996.

[205] J.L. Wear, *The use of NIR to predict residual ash in the IWTO-19 yield test – 2nd trial*, IWTO Report No. RWG 02, Barcelona, Spain, May 2002.

[206] J.L. Wear, *The use of NIR for predicting IWTO-19 residual ash in a commercial laboratory*, IWTO Report No. RWG 06, Nice, France, December 2002.

[207] J.L. Wear, *The use of NIR to predict residual ash in the IWTO-19 yield test*, IWTO Report No. RWG 02, Nice, France, November 2001.

[208] D.J. Petrie, J.J. Lidgard, J.W. Marler and A.H.M. Ireland, Wool Technol. Sheep Breed. 52(3) (2004) pp. 245–259.

[209] S.L. Ranford, R.J. Walls and H.E. Glassey, *1998 NIR residual grease round trials on scoured wool*, IWTO Report No. 20, Dresden, Germany, June 1998.

[210] S.L. Ranford, R.J. Walls and H.E. Glassey, *Dichloromethane extractable matter round trials – Soxhlet apparatus*, IWTO Report No. 21, Dresden, Germany, June 1998.

[211] J. Anson, *The identification and separation of vegetable matter types in core samples*, IWTO Report No. 3, Paris, France, January 1985.

[212] P.J. Connell, Wool Technol. Sheep Breed. 24(4) (1976) pp. 5–9.

[213] O.D. Wilkens and K.J. Whiteley, Wool Technol. Sheep Breed. 25(1) (1977) pp. 11–13.

[214] J. Anson, Wool Technol. Sheep Breed. 33(3) (1985) pp. 97–102.

[215] Anonymous, *Method for the determination of wool base and vegetable matter base, of core samples of raw wool (original version)*, IWTO 1930 DFT, IWTO-19-85(E), March 1985.

[216] I.M. Rogan, Wool Technol. Sheep Breed. 36(4) (1988/1989) pp. 126–135.

[217] P.R. Lamb, K.A. Hansford, D. Charlton and R.A. Rottenbury, Text. Asia 25(6) (1994) p. 36.

[218] E.M. Roberts, Wool Technol. Sheep Breed. 17(2) (1970) pp. 27–30.

[219] A.A. Dunlop and P.R. McMahon, Australian J. Agric. Res. 25 (1974) pp. 167–181.

[220] N. Jackson and A.M. Downes, Australian J. Agric. Res. 30 (1970) pp. 163–171.

[221] K.J. Whiteley, *Fibre diameter variation in wool*, Obj. Measurement Symp., Australia Paper No. 16, Melbourne, 1974.

[222] P. Maher, *Fibre diameter variation-measurement and importance in fine wool selection programmes*, Wool Annual, 1992, pp. 41–44.

[223] B. Quinnell, K.J. Whiteley and E.M. Roberts, *Variation in diameter of wool fibres – A Review*, Objective measurement of wool in Australia – technical report of the Australian Wool Board's Objective Measurement Policy Committee, AWC, Melbourne, Australia, October 1973, pp. 4.2–4.20.

[224] M. Brims, Wool Rec. 161(3694) (2002) pp. 62–63.

[225] X. Wang, *Diameter variation within single wool fibres and its effect on fibre properties*, in *Proceedings of the 11th Int Wool Text Res Conference*, CDRom, Leeds, United Kingdom, 2005.

[226] A.H. McKinley, P.A. Irvine, E.M. Roberts and M.W. Andrews, Proc. Aust. Soc. Anim. Prod. 11 (1976) pp. 181–184.

[227] B.P. Baxter, Wool Technol. Sheep Breed. 49(2) (2001) pp. 98–109.

[228] P.R. McMahon, *"Wool quality specifications in an integrated production and commerce manufacture system"*, in *Proceedings of the 5th Int. Wool Text. Res. Conference*, 4, Aachen, 1975, pp. 1–7.

[229] K.A. Hansford, Wool Technol. Sheep Breed. 40(1) (1992) pp. 2–9.

[230] K.A. Hansford, *On-farm ramifications of increased raw wool specification – Fibre diameter distribution,* Working paper presented at the CSIRO FDD Task Force Workshop held at CSIRO Div. of Wool Techn., Geelong (November 1991) and part of Appendix 2 in CSIRO Div. of wool Techn. Rep., G72, 1992.

[231] L. Yu, Xibei Fangzhi Gong Xuebao (Chinese), 9/1 (1995) pp. 1–6.

[232] S.A.S. Douglas, Wool Technol. Sheep Breed. 15(2) (1968) pp. 41–44.

[233] J.N. Skinner, Wool Technol. Sheep Breed. 7(2) (1960) pp. 41–44.

[234] D.J. Ward, *The development of fibre diameter technologies,* A presentation to the IWTO Technical and Standards Committee, Boston, USA, May 1997.

[235] J.F.P. James and A.E. Stearn, Wool Technol. Sheep Breed. 18(2) (1971) pp. 11–16.

[236] N. Jackson and H. Engel, Wool Technol. Sheep Breed. 28(3) (1980) p. 31.

[237] N. Jackson and H. Engel, Wool Technol. Sheep Breed. 28(3) (1980) pp. 32–37.

[238] R.N. Andrews and H. Hawker, Wool Technol. Sheep Breed. 30(1) (1982) pp. 32–35.

[239] J.W. Marler, *Method of determining fibre diameter distribution parameters and percentage of medullated fibres in wool and other animal fibres by the projection microscope,* IWTO Sliver Group Appendix 3, Boston, USA, May 1997.

[240] A.M. Downes, Wool Technol. Sheep Breed. 20(2) (1973) pp. 29–34.

[241] N. Jackson and A.M. Downes, Wool Technol. Sheep Breed. 22(2) (1975) p. 15.

[242] N. Jackson and A.M. Downes, Wool Technol. Sheep Breed. 22(2) (1975) pp. 38–43.

[243] L.J. Lynch and N.R. Michie, Wool Technol. Sheep Breed. 20(2) (1973) pp. 23–27.

[244] J.P. Connell and H.G. David, Wool Technol. Sheep Breed. 19(2) (1972) pp. 47–49.

[245] I.R. Edmonds, J. Text. Inst. 60 (1969) pp. 161–164.

[246] R.B. Whan and J.R. Paynter, J. Text. Inst. 58 (1967) pp. 273–278.

[247] J.A. Matthew, J. Text. Inst. 23 (1932) pp. T55–T70.

[248] M.C. Boshoff and P.J. Kruger, Text. Res. J. 41 (1971) pp. 573–576.

[249] Anonymous, *Using Optical Diffraction to Measure Fibre Fineness,* AWTA Ltd. Newsletter, Victoria, Australia, February 2003, pp. 24–26.

[250] J. Ewles, J. Text. Sci. (1928) pp. 101–102.

[251] J.A. Mathew, J. Text. Inst. 23 (1932) pp. T55–T70.

[252] W.J. Onions and B. Ellingham, British J. Appl. Phys. 10 (1959) pp. 328–332.

[253] V.V. Vakovlev, *Measurement of fiber and filament diameters by diffraction methods,* IVUZ, Teklnol. Tekstlnol Pram., No. 2, 1965, pp. 128–131: Technol. Textile Ind, USSR, No. 2, 1965, pp. 27–30.

[254] H.F.M. Van Schie, J.W. Marler and L.J.H. Barry, *Measurement of fibre diameter by image analysis,* in *Proceedings of the IWTO Tech. Comm. Meeting,* Report No. 9, Dubrovnik, June 1990. Also published in Wool Tech. Sheep Breed. 38(4) (1990) pp. 96–100.

[255] B.P. Baxter, M.A. Brims and D.C. Teasdale, Wool Technol. Sheep Breed. 40(4) (1992) pp. 131–134.

[256] G.J. Higgerson and K.J. Whiteley, Wool Technol. Sheep Breed. 31(2) (1983) pp. 65–67.

[257] J.W. Marler and H. McNally, Wool Technol. Sheep Breed. 35(4) (1987) pp. 194–196.

[258] M.J. Keogh and E.M. Roberts, Wool Technol. Sheep Breed. 33(3) (1985) pp. 108–112.

[259] K. Baird, and R.G. Barry, *Evaluation of the Sirolan Laserscan instrument. Part 1. Test Specimen preparation factors that influence the measured mean fibre diameter,* IWTO Report No. 5, Nice, France, December 1992.

[260] Anonymous, *The Airflow instrument: Using porosity to measure mean fibre diameter,* AWTA Ltd. Newsletter, Victoria, Australia, March 2004, pp. 18–21.

[261] P.J. Sommerville, Wool Technol. Sheep Breed. 48(2) (2000) pp. 102–146.

[262] B.P. Baxter, *Influences on comparisons between the mean fibre diameter of wools measured by Airflow and by projected image methods,* IWTO Report No. 13, Istanbul, Turkey, May 1993.

[263] B.T. Doak and T.J. Mahar, Wool Technol. Sheep Breed. 47(4) (1999) pp. 266–280.

[264] A.R. Lindsay and J.W. Marler, Wool Technol. Sheep Breed. 47(4) (1999) pp. 281–301.

[265] A.R. Lindsay and J.W. Marler, *The importance of the coefficient of variation of fibre diameter of blends of wool tops when measured by Airflow, Laserscan and OFDA,* IWTO Report No. 13, Boston, USA, May 1997.

[266] IWTO Test Method, IWTO-28-93, *Determination by the Airflow method of the mean fibre diameter of core samples of raw wool,* 1993.

[267] Australian Wool Corporation, *Report on Trials Evaluating Additional Measurement (TEAM) 1981–1988*, Australian Wool Corporation, Melbourne, 1988.

[268] M.A. Jackson, J.W. Marler and P.D. Morgan, *Analysis of commercial core/comb comparisons of mean fibre diameter by the Airflow method*, IWTO Report No. 6, Nice, France, December 1989.

[269] K. Baird, R.G. Barry and J.W. Marler, *Mean fibre diameter measurement differences using Airflow, projection microscope and Sirolan-Laserscan*, IWTO Report No. 8, Nice, France, December 1993.

[270] M.W. Andrews and P.A. Irvine, *The relationship between airflow and Projection microscope diameters*, IWTO Report No. 11, Nice, France, December 1972.

[271] P.J. Sommerville, *The effect of between fibre coefficient of variation on the fibre fineness measured by the Airflow*, IWTO Report No. 14, Boston, USA, May 1997.

[272] K.T. Fell, M.W. Andrews and J.F.P. James, J. Text. Inst. 63 (1972) pp. 125–129.

[273] R. Edmunds, *Effects of standard deviation on airflow measurements of mean fibre diameter*, IWTO Report, No. 5, Nice, France, November 1993.

[274] P.J. Sommerville, Wool Technol. Sheep Breed. 48(1) (2000) pp. 39–53.

[275] J.F.P. James and H.G. David, J. Text. Inst. 59 (1968) pp. 585–592.

[276] S.L. Anderson and F.L. Warburton, J. Text. Inst. 40 (1949) pp. T749–T758.

[277] M.W. Andrews and M.R. Bow, *The effect of using Airflow diameter in Martindale's equation for yarn evenness*, IWTO Report No. 3, Monaco, June 1980.

[278] N.F. Roberts, IWTO Report No. 9, June 1959.

[279] N.F. Roberts, *Remarque sur l'influence du CV de diamétre dans l'étalonnage d'un appareil Airflow*, IWTO Report No. 9, June 1972.

[280] P.J. Sommerville and D.C. Teasdale, *Introduction of SIROLAN-LASERSCAN as the Standard Service for certification of mean fibre diameter by AWTA Ltd*, Commercial and Technical Implications, IWTO Report No. RWG 03, Florence, Italy, May 1999.

[281] IWTO-8-61 (E) *Method of determining fibre diameter and percentage of modulated fibres in wool and other animal fibres by the projection microscope*, 1961.

[282] N.McM. Browne and W.R. Hindson, J. Text. Inst. 73 (1982) pp. 196–199.

[283] P.J. Sommerville, Wool Technol. Sheep Breed. 48(3) (2002) pp. 198–232.

[284] J.W. Marler, *Technical Committee*, IWTO Raw Wool Group, Attachment 5, New Delhi, India, March 1994.

[285] S.L. Anderson and R.C. Palmer, J. Text. Inst. 42 (1951) pp. P114–P116.

[286] S.L. Anderson and F. Benson, J. Text. Inst. 44 (1953) pp. P98–P104.

[287] H. Wang, T. Mahar, X. Liu and X. Wang, Text. Res. J. 77 (2007) pp. 981–991.

[288] S.L. Anderson and F. Benson, *Fibre ellipticity and its effect on diameter measurement*, IWTO Report, Paris, France, January 1952.

[289] M. Bona, T. Belleli and G. Blankenburg, *Correlation of the fibre ellipticity, snippet length and embedding medium with the mean diameter of mohair and wool*, IWTO Report No. 3, Nice, France, December 1992.

[290] J.G. Downes, J. Text. Inst. 8 (1975) pp. 300–301.

[291] S.L. Anderson and R.C. Palmer, *The effect of non-circular cross-section on fibre diameter measurement of wool by the profile method*, in *Proceedings of the Tech Committee* IWTO 1 (1947) pp. 29–30.

[292] G. Blankenburg, J. Haknen, K.-H. Lehmann, H. Pilippen and P. Spiegelmacher, *The problem: Fibre 'diameter' measurement in profile*, IWTO Sliver Group Appendix 1, Nice, France, December 1997.

[293] G. Blankenburg, H. Pilippen, P. Spiegelmacher and J. Hahnen, *Correlation of the fibre ellipticity, snippet length and embedding medium with the mean diameter of mohair and wool*, IWTO Report No. 3, Nice, France, December 1992.

[294] G. Blankenburg, H.J. Henning, H. Pilippen and P. Spiegelmacher, *Influence of snippet length on the microprojection diameter values*, IWTO Sliver Group Report No. 17, New Delhi, India, 1994.

[295] AWTA Ltd, *Using photometry to measure mean fibre diameter*, (1950–1970) AWTA Ltd. Newsletter, Victoria, Australia, April 2005, pp. 23–28.

[296] Anonymous, *Using photometry to measure mean fibre diameter*, AWTA Ltd. Newsletter, January 2006, pp. 18–23.

[297] A.R. Edmunds, B.A. Perry and J. Bedford, J. Text. Inst. 64 (1973) pp. 681–687.

[298] L.J. Lynch and N.A. Michie, Text. Res. J. 46 (1976) pp. 653–660.
[299] B.J. Crook, L.R. Piper and O. Mayo, Proc. Australian Assoc. Anim. Breed. Genet. 11 (1995) pp. 227–223.
[300] H.G. David, Wool Technol. Sheep Breed. 39(1) (1991) pp. 19–23.
[301] T. Dabbs and M. Glass, *Discrimination in the wool Fibre Diameter Analyser (FDA)*, IWTO Report No. 11, Uruguay, 1992.
[302] W.H.M. Lunney and P.A. Irvine, Text. Res. J. 52 (1982) pp. 217–227.
[303] AWTA Ltd. Newsletter, 50th Anniversary Edition, October 2007.
[304] J.W. Marler and H.W.M. Lunney, Text. Res. J. 53 (1983) pp. 677–681.
[305] B. Thompson and D.C. Teasdale, *FFDA Round trial*, IWTO Report No. 3, Tokyo, 1984.
[306] B. Thompson and D.C. Teasdale, *Measuring wool fibre diameter with the FFDA*, in *Proceedings of the 7th Int. Wool Text. Res. Conference*, 2, 1985, p. 127.
[307] G.R.S. Naylor, Wool Technol. Sheep Breed. 43(3) (1995) pp. 254–255.
[308] D. Charlton, Wool Technol. Sheep Breed. 43(3) (1995) pp. 212–228.
[309] K. Baird and R.G. Barry, *Evaluation of the Sirolan Laserscan instrument. Part 1. Accuracy and precision of the measurement*, Technical Committee IWTO Report No. 6, Nice, France, December 1992.
[310] K. Baird and R.G. Barry, *Measurement of mean fibre diameter using SIROLAN LASERSCAN: Results of an International Round Trial*, IWTO Report No. 19, Istanbul, Turkey, May 1993.
[311] K. Baird, R.G. Barry and J.W. Marler, *Comparison of mean fibre diameter measurements by Airflow and Laserscan for a wide range of wool types*, IWTO Report No. 7, I, Nice, France, December 1994.
[312] K.J. Fu, Q.Y. Jaing, Y. Feng and L.S. Yang, Wool Text. J. 37(2) (2009) pp. 51–55.
[313] P.A. Irvine, M.R. Bow and H.F.M. Van Schie, *Calibration of Sirolan-Laserscan with interwoollabs standard tops*, IWTO Report No. 11, Nice, France, December 1992.
[314] M.R. Bow and H.F.M. Van Schie, *The Sirolan-laserscan round trials*, IWTO Report No. 12, Nice, France, December 1992.
[315] H. Harig, *Report of the 1995 IWTO round trial – Part 1: Raw wool*, IWTO Report No. 15, Harrogate, England, 1998.
[316] H. Harig, *Report on the 1995 round trial. Part II: Wool Tops*, IWTO Report No. 16, Harrogate, England, June 1995.
[317] IWTO-12-93, *Measurement of the mean and distribution of fibre diameter using the Sirolan-Laserscan fibre diameter analyser*, IWTO test Method, 1993, pp. 12–93.
[318] J.W. Marler, *Amendments to Include Checks in IWTO-12*, IWTO RWG Appendix No. 2, Christchurch, New Zealand, April 2000.
[319] I.M. Treloar, *Laserscan wool fibre diameter measurements – Comparison of snippet preparation techniques, (Part II – Woollen spun yarns)*, IWTO Report No. 2, Nice, France, December 1997.
[320] I.M. Treloar, *Laserscan wool fibre diameter measurements – comparison of snippet preparation techniques*, IWTO Report No. 9, Boston, USA, May 1997.
[321] B.V. Holcombe and M.E. Papazian, *A technique for sampling and preparation of Laserscan snippets from woven and knitted fabrics*, IWTO Report No. 6, Nice, France, 1995.
[322] H.F.M. Van Schie, J.W. Marler and L.J.L. Barry, Wool Technol. Sheep Breed. 38(3&4) (1990/1991) pp. 96–100.
[323] H. McNally and A.R. Edmunds, *FIDAM-an image analysis system for fibre diameter measurement*, IWTO Report No. 7, Rio De Janeiro, June 1987.
[324] A.R. Edmunds and H. McNally, *Preliminary performance data for FIDAM*, IWTO Report No. 8, Rio De Janeiro, June 1987.
[325] B.P. Baxter and D.C. Teasdale, *Suitability of interwoollabs I H tops for calibration of the OFDA*, IWTO Report No. 9, Nice, France, December 1992.
[326] B.P. Baxter, M.A. Brims and T.B. Taylor, *Description and performance of the Optical Fibre Diameter Analyser (OFDA)*, IWTO Report No. 8, Nice, France, December 1991.
[327] B.P. Baxter, M.A. Brims and T.B. Taylor, *Further studies on the performance of the OFDA*, IWTO Report No. 8, Nice, France, April 1992.
[328] B.P. Baxter, *OFDA round trials with greasy cores*, IWTO Report No. 8, Nice, France, December 1992.
[329] M.A. Brims, *Introducing OFDA4000: A new instrument for simultaneous measurement of fibre length and diameter in tops*, IWTO Report No. SG 02, Barcelona, Spain, May 2002.

[330] G. Blankenburg, H.J. Henning, P. Spiegelmacher and H. Philippen, Wool Technol. Sheep Breed. 42(2) (1994) pp. 160–175.

[331] Anonymous, *Using image analysis to measure fibre diameter*, AWTA Ltd. Newsletter, Victoria, Australia, April 2007, pp. 25–29.

[332] S.A.G. Peyer, Text. Asia 24(5) (1993) p. 80.

[333] IWTO-47-00, *Measurement of the mean and distribution of fibre diameter of wool using an optic fibre diameter analyser (OFDA)*, 2000.

[334] R.C. Couchman, *A review of and recommended changes to IWTO 47 – Measurement of the mean and fibre diameter distribution of wool using the optical Fibre Diameter Analyser (OFDA)*, IWTO Report No. RWG01, Edinburgh, Scotland, May 2007.

[335] IWTO-8-66, Microtome Test Method, 1966.

[336] S.I. Denney, T.J. Mahar and J.W. Marler, *The effect of snippet length and fibre relaxation on the mean fibre diameter, fibre diameter distribution and fibre curvature measured by LASERSCAN and OFDA*, IWTO Report No. SG 03, Christchurch, New Zealand, April 2000.

[337] A.R. Edmunds, IWTO Technology and Standard Committee, Raw Wool Group Appendix 5, Cape Town Meeting, April 1996.

[338] D.W.F. Turpie, *Report on round trial no. 1 of the working group on "Objectionable" medullated fibres*, IWTO Technology and Standards Committee, Nice, December 1995.

[339] J.W. Marler, K.A. Hansford and I.M. McLachlan, Wool Technol. Sheep Breed. 50(4) (2002) pp. 832–839.

[340] M.A. Brims, *New OFDA developments and the use of the OFDA as a projection microscope*, IWTO Report No. 22, Istanbul, Turkey, May 1993.

[341] B. Xu and Y.L. Ting, J. Text. Inst. 87(Part 1 No. 2) (1996) pp. 274–283.

[342] B. Xu and Y.L. Ting, J. Text. Inst. 87(Part 2 No. 2) (1996) pp. 284–297.

[343] K.J. Whiteley, Wool Technol. Sheep Breed. 19(2) (1972) pp. 31–32.

[344] G.R.S. Naylor, Wool Technol. Sheep Breed. 46(2) (1998) pp. 144–153.

[345] D.G. Phillips, Wool Technol. Sheep Breed. 40(1) (1992) pp. 35–39.

[346] R.A. Rottenbury, Wool Technol. Sheep Breed. 40(1) (1992) p. 1.

[347] G.R.S. Naylor, Wool Technol. Sheep Breed. 45(4) (1997) pp. 221–226.

[348] J. Crook, Proc. Australian Assoc. Anim. Breed. Genet. 10 (1992) pp. 118–121.

[349] G.R.S. Naylor, Wool Technol. Sheep Breed. 40(2) (1992) pp. 40–43.

[350] G.R.S. Naylor, *Report on an international round trial to establish precision data on the Working Group Draft Test Method 'Measuring the Diameter Distribution of Fibre Ends in Sliver*, IWTO Report No. SG 01, Nice, France, 2001.

[351] T.J. Mahar and J.M.A. O'Keefe, Wool Technol. Sheep Breed. 50(4) (2002) pp. 826–831.

[352] G.R.S. Naylor, Wool Tech. Sheep Breed. 40(2) (1992) pp. 44–46.

[353] A.M. Van Zyl, *The relationship between Airflow and OFDA measurements for South African greasy wool and commercial mill tops*, IWTO Report No. SG 02, Christchurch, New Zealand, April/May 2000.

[354] Independent Laboratories Round Trial (ILRT) Group, *The effect of relaxation on the mean fibre diameter and fibre diameter distribution measured by LASERSCAN and OFDA*, IWTO Report No. SG 04, Christchurch, New Zealand, April 2000.

[355] P.J. Sommerville, *Measurement of the fineness of superfine wool: A comparison of Airflow, Laserscan and OFDA*, IWTO Report No. 15, Boston, USA, May 1997.

[356] D.G. Knowles, P.R. Greatorex and G.V. Barker, *Comparison of IWTO Test Methods for fibre diameter measurement of the New Zealand wool clip – Part 1: Mean fibre diameter*, IWTO Report No. 12, Dresden, Holland, June 1998.

[357] D.G. Knowles, P.R. Greatorex and G.V. Barker, *Comparison of IWTO Test Methods for fibre diameter measurement of the New Zealand wool clip – Part 2: Fibre diameter variability*, IWTO Report No. 13, Dresden, Holland, June 1998.

[358] P.A. Irvine and R.G. Barry, *An improved calibration model for the Sirolan-Laserscan*, IWTO Report No. 1, Nice, France, December 1997.

[359] I.M. Treloar, *Time intervals for achieving moisture equilibrium of wool samples in the standard atmosphere for conditioning*, IWTO Report No. SG 03, Florence, Italy, May 1999.

[360] P.J. Sommerville, *Measurement of the fineness of superfine wool: Effect of the revised LASERSCAN calibration function on comparisons between Airflow, LASERSCAN and OFDA*, IWTO Report CTF 04, Nice, France, December 1998.

[361] B.P. Baxter, Wool and Woollens India 35(4) (1998) pp. 27–45.

[362] J.W. Marler and H. Harig, *A comparison of diameter measurement technologies from Inter-woollabs international round trials for wool tops*, IWTO Report No. 10, Dresden, Germany, June 1998.

[363] B.P. Baxter, *Examination of the differences between OFDA and Laserscan on ultrafine wools*, IWTO Report No. 6, Nice, France, December 1997.

[364] M. Feughelman, J. Soc. Cosmet. Chem. 33 (1982) pp. 385–406.

[365] B.P. Baxter, Wool Technol. Sheep Breed. 45(4) (1997) pp. 267–282.

[366] W.A. Heath, J.W. Barkhuizen and O.E. Wright, *The relationship between mean fibre diameter measurements by Airflow and Laserscan for South African Wools*, IWTO Report No. RWG 03, Cairo, Egypt, May 2006.

[367] P. Baxter, *An initial investigation into the effects of medullation on comparisons between Airflow, OFDA and Laserscan*, IWTO Report No. 16, Dresden, Germany, June 1998.

[368] D.W. Turpie, *Objectionable medullated fibres in wool and mohair – results of round trial no. 2 and comparative results obtained by projection microscope*, IWTO Technology and Standard Committee, Sliver Group Appendix B, Nice meeting, December 1996.

[369] B. Thompson and D.C. Teasdale, *The length of snippets for FFDA measurement*, IWTO Technical Committee, Report No. 06, Ostende Meeting, June 1986.

[370] A.R. Edmunds, *Effects of snippet length on profile fibre diameter measurements*, IWTO Technical Committee, Report No. 14, Nice Meeting, December 1992.

[371] A.R. Edmunds, IWTO Technical Committee, Annex to the Minutes, New Delhi Meeting, March 1994.

[372] A.R. Edmunds and R.L. Ranford, *Conditioning effects on fibre diameter and curvature of the 12th series of IH tops*, IWTO Technology and Standard Committee, Raw Wool Group, Report No. 3, Nice Meeting, November/December 1999.

[373] V.E. Fish, T.J. Mahar and B.J. Crooke, *Fibre curvature morphometry and measurement*, IWTO Technology and Standard Committee, CFT 01, Nice Meeting, November/December 1999.

[374] H.W.M. Lunney and P.A. Irvine, Text. Res. J. 49(7) (1979) pp. 371–379.

[375] K. Baird, R.G. Barry, and J.W. Marler, *Comparison of mean fibre diameter measurements by Airflow and Laserscan for a wide range of wool types*, IWTO Technical Committee, Raw Wool Group, Report No. 2, Nice Meeting, 1994.

[376] T.P. Dabss, H. van Schie, and M. Glass, *The effect of fibre curvature on LASERSCAN diameter measures*, IWTO Technical Committee, Report No. 2, Nice Meeting, December 1994.

[377] A.D. Peterson, S.G. Gheradi and M.R. Ellis, Asian-Aust. J. Anim. Sci. A (2000) pp. 469–472.

[378] D.W.F. Turpie, *Report to the Chairman of the Sliver Group from the Chairman of the Working Group on the effects of Pre-treatment on Fibre Fineness*, IWTO Technology and Standard Committee, Sliver Group, Nice, December 1996.

[379] B.P. Baxter, *IWTO Sliver Group- Boston, May 1997 Some notes on the effects of relaxation on mean fibre diameter measurement*, IWTO Technology and Standard Committee, Sliver Group Appendix 1, Boston, May 1997.

[380] D.W. Turpie, *Report of results of round trial on relaxation and some related investigations*, IWTO Sliver Group Appendix 5, Nice, France, December 1996.

[381] B.P. Baxter and M.A. Brims, *Linearity – and focus independence of the OFDA*, IWTO Report No. 10, Nice, France, December 1992.

[382] M.R. Bow, H.F.M. Van Schie and P.A. Irvine, *An evaluation of some aspects of the Sirolan-Laserscan*, IWTO Report No. 24, Istanbul, Turkey, May 1993.

[383] InterWoollabs Management Committee, *Interwoollabs report on 1999 round trials*, IWTO Report No. SG 01, Christchurch, New Zealand, April 2000.

[384] D. Knowles and J.W. Marler, *The relationship between Mean Fibre Diameter Measurement By Airflow, LASERSCAN and OFDA for Australian Wool Tops*, IWTO Technical Committee, Raw Wool Group Report No. RWG 02, Florence Meeting, May 1999.

[385] D.W.F. Turpie, *Changes in dimensions of calibration samples snippets with changes in pre – treatment*, IWTO Technical Committee, Report No. 16, Cape Town Meeting, April 1996.

[386] IWTO Technology and Standards Committee, Sliver Group Minutes, Nice meeting, December 1996.

[387] IWTO Technology and Standards Committee, Sliver Group Minutes, Boston Meeting, May 1997.

[388] B.P. Baxter, *Some notes on the effects of relaxation on mean fibre diameter measurement*, IWTO Sliver Group Appendix 1, Boston, USA, May 1997.

[389] B.P Baxter, *Sample preparation effects on diameter – Notes from a brief 30 years history*, IWTO Relaxation Working Group Meeting, Nice, France, December 1996.

[390] H. Wang, X. Liu, T.J. Mahar and X. Wang, Int. J. Sheep Wool Sci. 54 (2006) pp. 1–17.

[391] A.R. Edmunds, *Relaxation, curvature and snippet length effects in profile fibre diameter measurement*, IWTO Report No. 20, Boston, USA, May 1997.

[392] M.A. Brims, *The theoretical comparison of OFDA and FDA class instruments for the measurement of fibre diameter distribution*, IWTO Report No. 21, Istanbul, Turkey, May 1993.

[393] P.J. Sommerville, *Fundamental principles of fibre fineness measurement: The Airflow Instrument*, IWTO Report No. CTF 03, Nice, France, December 1998.

[394] V.E. Fish and T.J. Mahar, *A modified technique for measurement of mean fibre diameter of yarns extracted from fabrics, using either OFDA100 or Laserscan*, IWTO Report No. PG 01, Frankfurt, Germany, June 2009.

[395] D.W. Crowe, V.E. Fish, J.W. Marler and T.J. Mahar, *A progress report on investigations into alternate test procedures for the measurement of fibre diameter on yarns and yarns drawn from fabrics*, IWTO Report No. PG 02, Edinburgh, Scotland, 2007.

[396] V.E. Fish, H.M. Wang, R.M.L. Hunter and T.J. Mahar, *A modified technique for measurement of mean fibre diameter of yarns extracted from fabrics*, IWTO Report No. PG 01, Beijing, China, 2008.

[397] D.W. Crowe and J.W. Marler, *Calibrating Laserscan for raw wool measurement and an evaluation of an alternative transport fluid*, IWTO Report No. RWG 03, Nice, France, 2000.

[398] G.R.S. Naylor, D.G. Phillips and C.J. Veitch, Wool Technol. Sheep Breed. 43(1) (1995) pp. 69–82.

[399] B.P. Baxter and D.J. Cottle, Wool Technol. Sheep Breed. 46 (2) (1998) pp. 154–171.

[400] K.J. Whiteley, B. Thompson, J.H. Standton and S.J. Welshman, Text. Res. J. 54 (1984) pp. 459–462.

[401] H.G. David, *The relation between standard deviation of diameter and mean diameter for commercial wool tops*, IWTO Report No. 4, Paris, January 1984.

[402] M.R. Bow and H.G. David, Wool Technol. Sheep Breed. 40(2) (1992) pp. 56–59.

[403] R. Ott, Bull. Inst. Text. France 77 (1958) p. 63.

[404] G.R.S. Naylor, J. Text. Inst. 87(2) (1996) pp. 265–273.

[405] H.G. David, *The coefficient of variation of fibre diameter of commercial wool tops – A Re-Examination of Ott's Data*, IWTO Report No. 4, Monaco, June 1975.

[406] M.R. Fleet, P.A. Irvine, R.W. Ponzoni and M.R. Bow, Wool Technol. Sheep Breed. 30(3) (1982) pp. 123–147.

[407] K.J. Whiteley, B. Thompson, J.H. Standton and S.J. Welsemen, *The distribution of fibre diameter in greasy wool sale lots*, IWTO Report No. 14, Venice, May 1982.

[408] J.F.P. James and H.G. David, J. Text. Inst. 60 (1969) p. 167.

[409] A.J. Ogden, J. Text. Inst. 60 (1969) pp. 164–166.

[410] C.J. Lupton, Sheep Goat Res. J. 11(3) (1995) pp. 111–121.

[411] B.P. Baxter and D.J. Cottle, *Fibre diameter distribution characteristics of midside (fleece) samples and their use in sheep breeding*, IWTO Report No. 12, Boston, USA, May 1997.

[412] M.R. Bow, L.M. Mooy and E. Zochowski, Wool Technol. Sheep Breed. 37(3) (1989) pp. 63–72.

[413] G.R.S. Naylor, *Fibre diameter distribution in sale lots and tops*, in *Proceedings of the Top-Tech '96 Conference*, Geelong, November 1996, pp. 277–283.

[414] M.R. Bow and K.A. Hansford, *Fibre diameter variability*, CSIRO Woolspec Seminar, 1994.

[415] K.L. Butler and M. Dolling, Wool Technol. Sheep Breed. 50(4) (2002) pp. 626–631.

[416] ASTM-D-3991, American Standard Specifications.

[417] G.R.S. Naylor, *The coarse fibre component and fibre end diameter characteristics of Australian wool tops*, IWTO Report No. 8, Harrogate, England, June 1995.

[418] P. Kenins, Wool Technol. Sheep Breed. 40(1) (1992) pp. 19–24.

[419] R.J. Mayfield, Text. Horizons 7(11) (1987) pp. 35–36.

[420] R.K. Garnsworthy, R.J. Mayfield, R.L. Gully, R.A. Westerman and P. Kenins, *Mechanics in cutaneous sensations of prickle and itch evoked by fabrics*, in *Proceedings of the 7th Int. Wool Text. Res. Conference*, 3, Tokyo, 1985, pp. 190–199.

[421] R.K. Garnsworthy, R.L. Gully. P. Kenins, R.J. Mayfield and R.A. Westerman, J. Neurophysiol. 59(4) (1988) pp. 1083–1097.

[422] Anonymous, Monash Review, 3 (1987) p. 87.

[423] R.A. Westermann, R.K. Garnsworthy, A. Walker, P. Kenins, R.L. Gully and P. Fergin, *Aspects of human cutaneous small nerve function: Sensations of prickle and itch*, 29th IUPS Satellite Symp., Budapest, 1984.

[424] C.J. Veitch and G.R.S. Naylor, Wool Technol. Sheep Breed. 40(1) (1992) pp. 31–34.

[425] CSIRO Division of Wool Technology, *Prickle and its prevention*, Textile Asia, 19(5) (1988) pp. 95–99.

[426] J.E. Smith, Textiles 22(1) (1993) pp. 18–20.

[427] G.R.S. Naylor, C.J. Veitch, R.J. Mayfield and R. Kettlewell, Text. Res. J. 62(8) (1992) pp. 487–493.

[428] C.J. Veitch, P. Kenins and G.R.S. Naylor, *Comparison between laboratory protocol and simulated wear conditions for the evaluation of prickle*, in *Proceedings of the 8th Int. Wool Text. Res. Conference*, Christchurch, 5 (1990) pp. v286–v295.

[429] M. Matsudaira, J.D. Watt and G.A. Carnaby, J. Text. Inst. 81(3) (1990) pp. 288–299.

[430] Y. Li, Text. Prog. 31(1/2) (2001) pp. 1–37, 99–101.

[431] D.G. Knowles, *Discrepancies in the calculation of coarse edge fibre content between Laserscan and OFDA*, Appendix 3, IWTO Raw Wool Group Meeting, Dresden, June 1998.

[432] J.W. Marler, *Definition of comfort factor-recommended inclusions to IWTO-12 and IWTO-47*, Appendix 1, IWTO Raw Wool Group Meeting, Nice, November 1999.

[433] A.F. Botha and L. Hunter, South African J. Sci. 103(3/4) (2007) pp. 174–176.

[434] K.J. Whiteley and B. Thompson, Text. Res. J. 55 (1985) pp. 107–112.

[435] G.V. Barker and S.J. Fookes, *A collaborative study on the application of measurement technology to New Zealand fine wool*, IWTO Report No. 19, Boston, USA, May 1997.

[436] G.R.S. Naylor and D.G. Phillips, *Skin comfort of wool fabrics*, in *Proceedings of the 9th Int. Wool Text. Res. Conference* II, Biella, 1995, pp. 203–209.

[437] B.H. Mackay, Wool Technol. Sheep Breed. 15(2) (1968) pp. 22–30.

[438] G.R.S. Naylor, Wool Technol. Sheep Breed. 50(4) (2002) pp. 698–704.

[439] T.J. Mahar, J.M.A. O'Keefe and R.C. Couchman, *A preliminary report on the use of raw wool measurements to predict fibre end characteristics of top*, IWTO Report No. SG 01, Barcelona, Spain, May 2002.

[440] D.W.F. Turpie, *The processing characteristics of South African wools Part 10 – The influence of relatively large variations in diameter on the processing performance of South African long wools up to spinning*, SAWTRI Techn. Report No. 303, May 1976.

[441] L. Gallico, Ann. Sci. Text. Belges 2 (1971) p. 36.

[442] H.J. Henning and D.R. Merx, Ann. Sci. Text. Belges 4 (1971) p. 22.

[443] D. Charlton, J.R. Eley and R.A. Rottenbury, *Processing studies on clip preparation: Three clips heavily contaminated by vegetable matter*, IWTO Report No. 1, Tokyo, May 1984 and Wool Techn. & Sheep Breed. 33(1) (1985) pp. 15–24.

[444] P.J.M. Bell, Wool Technol. Sheep Bred. 35(2) (1987) pp. 101–108.

[445] R.A. Rottenbury, M.W. Andrews and G.H. Brown, *The association between worsted processing performance and raw wool characteristics*, IWTO Report No. 4, Christchurch, New Zealand, 1981.

[446] W. Von Bergen and J.H. Wakelin, Text. Res. J. 22(2) (1952) pp. 123–137.

[447] K.O.A. Kurdo, K.J. Whiteley and L.J. Smith, *The influence of resistance to compression on the processing performance of superfine wools. Part 1: Topmaking*, IWTO Report No 12, Tokyo, 1984.

[448] G.R. Stanbury and H.E. Daniels, J. Text. Inst. 28 (1937) pp. T188–T201.

[449] P.P. Townend, A. Hewitt and C.K. Chu, *The comparative contribution of some card variables to the formation of neps, using metallic clothing*, in *Proceedings of the 5th Int. Wool Text. Res. Conference*, IV, Aachen, 1975, pp. 177–188.

[450] G. Blankenburg and G. Satlow, J. Text. Inst. 53 (1962) pp. T489–T491.

[451] D.A. Ross, J. Text. Inst. 51 (1960) pp. T240–T259.

[452] E.M. Pohle, D.D. Johnston, H.D. Ray, W.J. Manning and W.A. Mueller, Text. Res. J. 23 (1953) pp. 663–671.

[453] H.G. David and M.W. Andrews, *The change in fibre diameter distribution of wool during early-stage processing*, IWTO Report No. 7, Tokyo, May 1984.

[454] D.W.F. Turpie and E. Gee, *The properties and performances during topmaking and spinning of a wide range of South African wools*, in *Proceedings of the 6th Int. Wool Text. Res. Conference*, Pretoria, 3, 1980, pp. 293–311.

[455] D.W.F. Turpie, *The processing characteristics of SA wools. Part 9: The influence of limited variations in both length and diameter on the processing performance of mixtures of SA wools up to spinning*, SAWTRI Techn. Report No. 284, March 1976.
[456] J. Menkart and J.G. Detenbeck, Text. Res. J. 27(9) (1957) pp. 665–689.
[457] W. von Bergen, *Wool Handbook*, Interscience, John Wiley, 1963, p. 171.
[458] J. Menkart and B. Joseph, Text. Res. J. 28 (1958) pp. 940–945.
[459] K. Lees, Wool Rec. 119 (1971) pp. 24–25.
[460] P.J.M. Bell, Private Communication.
[461] K. Butler, M. Dolling, D.J. Marland, G.R.S. Naylor, D.G. Phillips and C.J. Veitch, *The usefulness of midside sampling as a predictor of top fibre diameter distribution characteristics*, Working paper presented at the CSIRO FDD Task Force Workshop held at CSIRO Division of Wool Technology, Geelong, November 1991 and part of Appendix 2 in CSIRO Division of Wool Technology Report No. G72, 1992.
[462] A.W. Khandaker and W.R. Lang, Text. Res. J. 32(9) (1962) pp. 786–788.
[463] AWTA Ltd., *Processing of Classed Grower and Bulk Class Lots of Different Uniformity Index*, IWTO Report No. CTF 03, Nice, France, November 2001.
[464] M.W. Andrews, S.A.S. Douglas, A.H.M. Ireland, R.J. Quirk and R.A. Rottenbury, *Report to the Wool measurement research advisory committee of the Australian wool corporation*, IWTO Final TEAM Rep., IWTO Conference, Paris, January 1985.
[465] B.P. Baxter, *The effects of calibration errors on between-laboratory and between-instrument diameter comparisons*, IWTO Report No. SG01, November 1999.
[466] K.A. Hansford, J.W. Marler and I.M. McLachlan, Wool Technol. Sheep Breed. 50(2) (2002) pp. 140–167.
[467] D.G. Knowles, Wool Technol. Sheep Breed. 48(3) (2000) pp. 183–197.
[468] E.M. Roberts, Wool Technol. Sheep Breed. 2(1) (1955) pp. 33–45.
[469] K.J. Whiteley, Wool Technol. Sheep Breed. 3(1) (1956) pp. 71–72.
[470] E.M. Roberts, Wool Technol. Sheep Breed. 4(1) (1957) pp. 85–90.
[471] W.L. McGarry, Wool Technol. Sheep Breed. 7(1) (1960) pp. 115–116.
[472] J.C. Potter and A.D. Carey, Wool Technol. Sheep Breed. 8(1) (1961) pp. 139–141.
[473] B. McGuirk, Wool Technol. Sheep Breed. 26(1) (1978) pp. 17–22.
[474] J.W. Marler and B.C. Couchman, Wool Technol. Sheep Breed. 49(4) (2001) pp. 311–326.
[475] R.N. Andrews, J.T.J. Land and K.G. Dodds, Wool Technol. Sheep Breed. 45(1) (1997) pp. 35–50.
[476] R.I. Gillies and W. Stadler, Wool Technol. Sheep Breed. 45(1) (1997) pp. 27–34.
[477] L.R. Piper and J. Lax, Wool Technol. Sheep Breed. 40(1) (1992) pp. 10–13.
[478] P. Taylor and K. Atkins, Wool Technol. Sheep Breed. 40(3) (1992) pp. 97–99.
[479] R.W. Ponzoni and F.D. Brien, Wool Technol. Sheep Breed. 41(2) (1993) pp. 156–160.
[480] J.C. Greeff, Wool Technol. Sheep Breed. 50(1) (2002) pp. 11–15.
[481] R. Lewer and Y. Li, Wool Technol. Sheep Breed. 42(2) (1994) pp. 103–111.
[482] M. Yamin, P.I. Hind, R.W. Ponzoni, J.A. Hill, W.S. Pitchford and K.A. Hansford, Wool Technol. Sheep Breed. 47(3) (1999) pp. 151–158.
[483] P.R. Lamb, *Metrology and processing research: Implication for growers*, in *Proceedings of the Wool Industry Science Technology Conference*, CDRom, Hamilton, 2002, also Wool Technol. Sheep Breed. 50(4) (2002) pp. 671–678.
[484] K.D. Atkins and S.J. Semple, Wool Technol. Sheep Breed. 51(1) (2003) pp. 1–15.
[485] M.A. Brims, A.D. Peterson and S.G. Gherardi, *Introducing the OFDA2000 – For rapid measurement of diameter profile on greasy wool staples*, IWTO Report No. RWG 04, Florence, Italy, May 1999.
[486] B.P. Baxter, Wool Technol. Sheep Breed. 49(2) (2001) pp. 133–155.
[487] B.P. Baxter, *Precision of measurement of diameter and diameter-length profiles of greasy wool staples on-farm, using the OFDA2000 instrument*, Available at www.i-activewool.com.au
[488] M. Brims, *Along fibre diameter and cleanliness measurement using OFDA*, IWTO Report No. 23, Boston, USA, May 1997.
[489] M.A. Brims, A.D. Peterson and S.G. Gherardi, *Introducing the OFDA2000 – for rapid measurement of diameter profile on greasy wool staple*, IWTO Meeting, Report No. RWG 04, Florence, May 1999.
[490] B.P. Baxter, *Precision measurement of diameter, and diameter-length profile of greasy wool staples on-farm, using the OFDA2000 instrument*, in *Proceedings of the 10th Int Wool Text Res Conference*, CDrom, Aachen, Germany, 2000, pp. RW-2,1–RW-2,10.

[491] A.D. Peterson and S.G. Gherardi, Wool Technol. Sheep Breed. 49(2) (2001) pp. 110–132.

[492] R. Behrendt, K. Konstantinov, F. Brien, M. Ferguson and C. Gloag, Wool Technol. Sheep Breed. 50(4) (2002) pp. 780–786.

[493] F.D. Brien, Int. J. Sheep Wool Sci. 53(1) (2005) pp. 46–56.

[494] D.J. Brown, B.J. Crook and I.W. Purvis, Wool Technol. Sheep Breed. 48(2) (2000) pp. 86–93.

[495] K.A. Hansford, Wool Technol. Sheep Breed. 47(1) (1999) pp. 19–33.

[496] K.A. Hansford, J.W. Marler and I.M. Mclachlan, *A comparison of the performance of on-farm OFDA2000 and in-store FLEECESCAN. Part 2. Sheep selection and fleece classing*, IWTO Report No. CTF 09, Barcelona, Spain, May 2002.

[497] W. Humphries, L. Vuckovic, W.A. Aspros and G.J. Higgerson, *Sirolan Fleecescan: Performance and benefits of in-shed individual fleece testing*, IWTO Report No. CTF 04, Shanghai, China, May 2001.

[498] J.W. Marler, K.A. Hansford and I.M. McLachlan, *A comparison of the performance of the OFDA2000, FLEECESCAN and LASERSCAN – The precision of the measured fibre characteristics of a fleece*, IWTO Report No. CTF 08, Barcelona, Spain, May 2002.

[499] A.D. Peterson and S.G Gherardi, Wool Technol. Sheep Breed. 49(2) (2001) pp. 110–132.

[500] M.B. Ferguson, C.M. Gloag, R. Behrendt and F.D. Brien, Wool Technol. Sheep Breed. 50(4) (2002) pp. 787–791.

[501] J.W. Marler, *The impact of test result precision on on-farm decisions*, IWTO Report No. CTF 04, Nice, France, December 2001.

[502] K.A. Hansford, J.W. Marler and I.M. McLachlan, Wool Technol. Sheep Breed. 50(4) (2002) pp. 812–818.

[503] J.W. Marler and P. Baxter, Wool Technol. Sheep Breed. 52(1) (2004) pp. 43–96.

[504] P. Baxter and J.W. Marler, Wool Technol. Sheep Breed. 52(1) (2004) pp. 124–170 [IWTO Report No. CTG 02, Evian, France, May 2004].

[505] A.R. Edmunds, *OFDA measurement of additional fibre parameters: Some preliminary results*, IWTO Report No. 19, Harrogate, England, June 1995.

[506] P. Baxter and B. Johnston, *OFDA2000 proficiency trial*, IWTO Report No. CTF 01, Nice, France, November 2002.

[507] B.P. Baxter, Wool Technol. Sheep Breed. 49(1) (2001) pp. 42–52.

[508] D. Cottle, Wool Technol. Sheep Breed. 49(4) (2001) pp. 325–326.

[509] A. Peterson, Wool Technol. Sheep Breed. 49(4) (2001) pp. 324–325.

[510] P.D. Morgan, Wool Technol. Sheep Breed. 38(1) (1990) pp. 21–23.

[511] D.J. Cottle, C.D. Almeida, B.P. Baxter and D.J. Petrie, Wool Technol. Sheep Breed. 44(4) (1996) pp. 295–303.

[512] B.P. Baxter, Wool Technol. Sheep Breed. 49(2) (2001) pp. 133–155.

[513] J.W. Marler and D.W. Crowe, Wool Technol. Sheep Breed. 50(4) (2002) pp. 840–846.

[514] J.W. Marler, K.A. Hansford and I. McLachlan, *A comparison of the performance of on-farm OFDA2000 and in-store FLEECESCAN. Part 2. Sheep selection and fleece classing*, IWTO Report No. CTF 09, Barcelona, Spain, May 2002.

[515] J.W. Marler, K.A. Hansford and I. McLachlan, *A comparison of the performance of on-farm OFDA2000 and in-store FLEECESCAN – Part 1: The precision of the measured fibre characteristics of a fleece*, IWTO Report No. CTF 08, Nice, France, December 2002.

[516] J.W. Marler, Wool Technol. Sheep Breed. 49(4) (2001) pp. 284–301.

[517] K.D. Atkins, Int. J. Sheep Wool Sci. 53(2) (2005) pp. 43–46.

[518] R. House, R. Bilney, D. Ladyman, C.M. Oldham and M. Yelland, Proc. Aust. Soc. Anim. Prod. 24 (2002).

[519] A.D. Peterson, S.G. Gherardi and P.T. Doyle, Aust. J. Agric. Res. 49 (1998) pp. 1181–1186.

[520] A.D. Peterson, J.C. Greeff, C.M. Oldham, D.G. Masters and S.G. Gherardi, Proc. Asian Aust. Anim. Prod. 23A (2000) p. 469.

[521] A.D. Peterson and C.M. Oldham, *The influence of date of shearing on the processing performance to top of mini-commercial consignments of Merino fleece wools grown in either South Western or Eastern Australia – 2. Improved prediction from the FD profiles of staples from component sale lots*, in *Proceedings of the 10th Int Wool Text. Res Conference*, CDrom, Aachen, Germany, 2000, pp. RW-P3.1–RW-P3.8.

[522] C.M. Oldham, S.G. Gherardi, B.R. Paganoni and M.R. Yelland, Proc. Aust. Soc. Anim. Prod. 24 (2002).

[523] K.A. Hansford and J.P. Kennedy, *The relationship between variation in fibre diameter along staples and staple strength*, in Proceedings of the 8th Int. Wool Text. Res. Conference, I, Christchurch, New Zealand, 1990, pp. 590–598.

[524] A.D. Peterson, *The components of staple strength*, MSc Thesis, University of Western Australia, Nedlands, 1997.

[525] B.A. McGregor and P.C. Toland, Wool Technol. Sheep Breed. 50(4) (2002) pp. 819–825.

[526] P.R. Lamb, Wool Technol. Sheep Breed. 52(2) (2004) pp. 171–192.

[527] J.L. Smith, I.W. Purvis and M.G. Haigh, Int. J. Sheep Wool Sci. 53(2) (2005) pp. 1–20.

[528] K.A. Hansford, *Influence of nutrition and reproduction on length, strength and position of break of Merino wool*, PhD Thesis, The University of New South Wales, Sydney, 1989.

[529] J.L. Smith, I.W. Purvis and G.J. Lee, Int. J. Sheep Wool Sci. 54(2) (2006) pp. 54–61.

[530] A.D. Peterson, S.G. Gherardi and M.R. Ellis, Asian-Aust. J. Anim. Sci. 23 (2000).

[531] K.A. Hansford and J.P. Kennedy, Proc. Aust. Soc. Anim. Prod. 17 (1988) p. 415.

[532] C.M. Oldman, G.R.S. Naylor, A.D. Peterson, G.A. Robinson and A.N. Thompson, Asian-Aust. J. Anim. Sci. 20 (2000).

[533] P.R. Lamb, *The impact of CV(D) and crimp or curvature on hauteur and romaine*, IWTO Report No. CTF 02, Christchurch, New Zealand, 2000.

[534] K.A. Hansford, Wool Technol. Sheep Breed. 45(4) (1997) pp. 309–320.

[535] K.A. Hansford, *Modelling of fibre breakage*, in Proceedings of the Top-Tech '96', CSIRO Division of Wool Technology, 1996, pp. 33–35.

[536] P.J. Sommerville, Wool Res. 161(3694) (2002) pp. 60–61.

[537] K.A. Hansford, *A study of the specification and topmaking performance of Western Australian fleeces and sale lots*, IWTO Report No. 15, Nice, France, December 1997.

[538] K.A. Hansford, *Fleece specification and processing prediction*, in Proceedings of the Wool-spec. 94, Seminar 'Specification of Australian Wool and its Implication for Marketing and Processing', CSIRO Division of Wool Technology, Sydney, Australia, 1994, pp. J1–J21.

[539] S.G. Gherardi, *New technologies from fibre to fabric*, Western Australian Department of Agriculture Report. Available at http://comm.agric.wa.gov.au/programs/wool/conference_papers.htm

[540] M. Yelland, *Managing the fibre profile*, Western Australian Department of Agriculture Report. Available at http://comm.agric.wa.gov.au/programs/wool/conference_papers.htm

[541] C. Scrivener, *Controlling fibre diameter profile and staple strength*, The Mackinnon Project Newsletter, March 2001.

[542] P.J. Hilton and A.I. Gardiner, *Fourier diameter measurement*, in Proceedings of the 25th NELCON88 Conference, 25, 1988, pp. 171–175.

[543] D.J. Brown, B.J. Crook and I.W. Purvis, Wool Technol. Sheep Breed. 48(1) (2000) pp. 1–14.

[544] D.W.F. Turpie, M.A. Strydom and J. Cizek, *The automatic measurement of the length, strength and profile of raw wool staples and their relevance to textile processing*, in Proceedings of the 2nd World Merino Conference, 3, Maastricht, Nederland, 1986, pp. 234–243.

[545] J. Grignet, P. Delfosse, P. Godard, J. Lenoir, W.D. Ainsworth and K.P. Russel, *Use of a modified sensor to evaluate the length and variation in linear mass (cross-section) of raw wool staples*, IWTO Report No. 14, Copenhagen, Denmark, May 1983.

[546] K.A. Hansford, G.B. Emery and D.C. Teasdale, Wool Technol. Sheep Breed. 33(1) (1985) pp. 25–27.

[547] A.C. Schlink, S.K. Clark and A.M. Murray, Wool Technol. Sheep Breed. 47(2) (1990) pp. 118–127.

[548] A. Dircks, Wool Technol. Sheep Breed. 3(2) (1956) pp. 31–33.

[549] A.M. Ford, *The measurement of the diameter-length distribution of New Zealand wool*, in Proceedings of the 8th Int. Wool Text. Res. Conference, II, Christchurch, New Zealand, 1990, pp. 115–124.

[550] D.A. Ross, N. Z. J. Agric. Res. 4 (1961) pp. 57–71.

[551] D.J. Brown, A.C. Schlink and B.J. Crook, Wool Technol. Sheep Breed. 47 (1999) pp. 170–183.

[552] R.C. Couchman, *Defining Hauteur and top & sliver 'length' distribution naming conventions*, IWTO Report No. SG 01, Shanghai, China, November 2004.

[553] R.C. Couchman, *Modelling of length distribution in the OFDA4000*, IWTO Report No. SG 02, Shanghai, China, November 2004.

[554] J. Grignet, *Foundations and metrology of the Almeter wool fibre length measurement*, IWTO Report No. SG 02, Dubrovnik, Croatia, November 2003.

[555] R.C. Palmer, *Fibre length measurement: The present position*, IWTO Report No. 1, Paris, France, 1947.

[556] G.R. Burling-Claridge and E.J. Wood, *A card waste test for wool*, WRONZ Report, Christchurch, New Zealand, 1995.

[557] E.J. Wood, Wool Technol. Sheep Breed. 47(3) (1999) pp. 218–229.

[558] J. Wear and P. Baxter, *A survey of length distribution data for N Z scoured consignments that have been measured for length after carding (LAC)*, IWTO Report No. 15, Dresden, Germany, June 1998.

[559] B.P. Baxter and J.L. Wear, *Comparison of 3 methods of predicting combing performance from raw wool test results*, IWTO Report No. 5, Nice, France, December 1997.

[560] H.G. David, *The relationship between staple length and mean fibre length for wool*, IWTO Report No. 19, Nice, December 1992.

[561] V.E. Fish, T.J. Mahar, P.J. Semmel and J.M. Dempsey, *The examination of decrimped staple length to differentiate atypical wool types as a potential aid to purchasing and processing prediction*, IWTO Report No. CTF 03, Buenos Aires, Argentina, May 2003.

[562] J. Cizek and D.W.F. Turpie, *An introduction to the SAWTRI length/strength tester for raw wool staples*, SAWTRI Technical Report No. 561, 1985.

[563] D.W.F. Turpie and J. Cizek, *Rapid estimation of fibre length distributions in wool staples by means of information provided by the SAWTRI length/strength tester*, SAWTRI Tech. Rep. No. 596, 1987.

[564] D.W.F. Turpie, M.A. Strydom and J. Cizek, in *Proceedings of the 2nd World Merino Conference*, 2, 321, Madrid, 1986.

[565] J.E. Deurden and V. Bosman, Dept. Dir. Vet. Serv. Anim. Ind. 17 (1931) p. 789.

[566] Y. Litav and E. Alexander, *Mechanical properties of felted balls as a function of fibre crimp*, in *Proceedings of the 5th Int. Wool Res. Conference*, 3, Aachen, 1975, pp. 493–502.

[567] J.N. Doney, Aust. J. Agric. Res. 10(2) (1959) p. 299.

[568] W.R. Lang and A.M. Chaudhri, J. Text. Inst. 39 (1948) pp. T249–T252.

[569] B.F. Short, Proc. Aust. Soc. Anim. Prod. 5 (1964) p. 350.

[570] M.W. Andrews, H.G. David, J.P. Connell, P. Heuer, P.A. Irvine, H.W. Lunney, J.W. Marler and R.A. Rottenbury, CSIRO Sheep and Wool Research, Geelong, Australia, 1977–1978, p. 109.

[571] M.W. Andrews, R.A. Rottenbury and G.H. Brown, *Stratified sampling for single-fibre-length measurement of greasy wool*, IWTO Report No. 3, Paris, December 1975.

[572] P.J. Kruger, J. Text. Inst. 62 (1971) pp. 47–55.

[573] P.J. Murray, *The effect of nutrition on wool growth, fibre and skin characteristics and live-weight gain of broad and fine wool sheep within two Merino strains*, PhD thesis, University of Western Australia, Perth, 1996.

[574] P. Delfosse, J Grignet and J. Lenoir, *Prediction of fibre length distribution and noil on top sliver from raw wool measurements using computer simulation technique*, IWTO Report No. 5 Paris, France, January 1986.

[575] P.R. Lamb and K.M.S. Curtis, *The influence of staple strength, length, curvature and diameter distribution on price and processing performance*, IWTO Report No. CTF 01, Hobart, Australia, April 2005.

[576] A.C. Schlink, G. Mata and R.M. Lewis, Wool Technol. Sheep Breed. 46 (1998) pp. 271–285.

[577] A.C. Schlink, G. Mata, J.M. Lea and A.J.M. Ritchie, Aust. J. Exp. Agric. 39 (1999) p. 507.

[578] R.A. Rottenbury and M.W. Andrews, Wool Technol. Sheep Breed. 22(1) (1975) pp. 27–30.

[579] M.W. Andrews and R.A. Rottenbury, J. Text. Inst. 66 (1975) pp. 200–202.

[580] E. Gee, SAWTRI Bull. 9(2) (1975) pp. 26–33.

[581] M.R. Bow, *Staple length*, in *Proceedings of the seminar 'Staple Length and Staple Strength of Greasy Wool : Measurement and Effects on Worsted Processing'*, CSIRO Division of Textile Physics, Ryde, New South Wales, December 1979.

[582] A.C. Schlink, D.J. Brown and M. Longree, Wool Technol. Sheep Breed. 49(3) (2001) pp. 202–211.

[583] N.R. Adams and R.W. Kelly, Asian-Aust. J. Anim. Sci. 13(Suppl. C) (2000) pp. 20–29.

[584] A.N. Thompson, P.I. Hynd, A.D. Peterson and A.J.M. Ritchie, *Fibre strength and the proportion of discontinuous fibres in relation to staple strength in Merino sheep*, in *Proceedings of the 9th Wool Text. Res. Conference*, Vol. II, Biella, Italy, 1995, pp. 143–151.

[585] P.I. Hynd, A. Hughes, C.R. Earl and N.M. Penno, Aust. J. Agric. Res. 48 (1997) p. 1089.
[586] A.R. Bray, D.R. Scobie and J.L. Woods, *Genetic improvement in the wool strength of Romney sheep*, in *Proceedings of the 9th Int. Wool Text. Res. Conference*, Vol. II, Biella, Italy, 1995, pp. 173–181.
[587] A.C. Schlink, J.R. Briegel, J. Greeff, A.N. Thompson and N.R. Adams, Anim. Prod. Aust. 22 (1998) p. 421.
[588] B.F Short, Proc. Aust. Soc. Anim. Prod. 5 (1964) p. 350.
[589] A.C. Schlink and A.E. Dollin, Wool Technol. Sheep Breed. 43(4) (1995) pp. 268–284.
[590] A.N. Thompson, A.C. Schlink, A.D. Peterson and P.I Hynd, Aust. J. Agric. Res. 49 (1998) p. 1173.
[591] A.C. Schlink, M. Sanders and D.E. Hollis, Asian-Aust. J. Anim. Sci. 13(Suppl. A) (2000) p. 253.
[592] M.L. Ryder, Aust. J. Agric. Res. 18 (1967) p. 683.
[593] J. Fowler, *Components of fibre length variability in greasy wool*, 4th ed., Australian wool-classing, 1990, p. 64 and The Indian Text. J. 42 (1990).
[594] R.A. Rottenbury, M.R. Bow and D.J. Allen, *Staple length variation in merino locks*, IWTO Report No. 4, June 1979.
[595] D.W.F. Turpie, *The processing characteristics of SA wools. Part VI : The influence of style, length and class description on the processing performance up to combing of mixtures of South African wools*, SAWTRI Technical Report No. 250, June 1975.
[596] P.R. McMahon, Wool Technol. Sheep Breed. 23(1) (1976) pp. 8–10.
[597] V. Aldrich de, P.J. Kruger and D.W.F. Turpie, *The carding and combing of wools of different fibre lengths*, SAWTRI Technical Report No. 136, 1970.
[598] J.H. Wakelin, H.W. Lambert and D.J. Montgomery, Text. Res. J. 26(9) (1956) p. 665.
[599] Anonymous, Wool Research 1918–1954, 3, WIRA, Leeds, 1955.
[600] Anonymous, Wool Sci. Rev. No. 5 (1950) p. 22–31.
[601] K. Fujino and S. Kawabata, *Probabilistic and mathematical model of fibre assembly and its application to drafting process analysis using specially designed analogue computer <<spinning simulator>>*, in *Proceedings of the 3rd Int. Wool Text. Res. Conference*, 4, Cirtel, Paris, 1965, pp. 151–161.
[602] R. Bownass, *Changes in fibre length during early worsted processing*, IWTO Technical Committee – IWTO Report (raw wool length working group), Paris, January 1984.
[603] R.A. Rottenbury, M.W. Andrews, P.J.M. Bell and R. Bownass, J. Text. Inst. 77(3) (1986) p. 179 and IWTO Report No. 13, January 1985.
[604] P. Godard, J. Grignet and J. Brach, *Evaluation of fibre length distribution from raw wool to top and yarn*, in *Proceedings of the 5th Int. Wool Text. Res. Conference* IV, Aachen, Germany, 1975, pp. 107–114.
[605] D.J. Allen, *Variability of fibre length in tops*, IWTO Report No. 11, Nice, 1991.
[606] E. Gee, SAWTRI. Bulletin, 11(3) (1977) pp. 16–23.
[607] R.C. Couchman and J. Turk, *Review and re-Draft: IWTO17–04-Determination of Fibre Length & Distribution Parameters*, IWTO Report No. SG 01 Evian, France, May 2004.
[608] R.C. Couchman and P.J. Holmes, *Report on equivalence round trials for the AL2000*, IWTO Report No. SG 03, Dubrovnik, Croatia, November 2003.
[609] J. Grignet, Ann. Sci. Text. Belges 10 (1962) pp. 70–93.
[610] J. Grignet, Wool Sci. Rev. 56 (1980) pp. 81–99.
[611] R.C. Couchman and P. J. Holmes, *Recommended changes to IWTO 17 to accommodate equivalence of the AL2000*, IWTO Report No. SG 01, Istanbul, Turkey, November 2003.
[612] M.A. Brims, *Improving agreement on Hauteur measurement between OFDA4000 and Almeter*, IWTO Report No. SG 01, Shanghai, China, November 2004.
[613] Ch. Bernardin and J. Grignet, Wool Technol. Sheep Breed. 42(1) (1994) pp. 51–57.
[614] P. Baxter and M. Brims, *Performance of OFDA4000 instruments in three mills laboratories*, IWTO Report No. SG 01, Beijing, China, April 2008.
[615] B.P. Baxter, *Report on the 2004/05 OFDA4000 and Almeter/OFDA100/Laserscan round trial*, IWTO Report No. SG 02, Hobart, Australia, April 2005.
[616] M.A. Brims, *Progress of the OFDA4000 and comparison with Almeter*, IWTO Report No. SG 02, Buenos Aires, Argentina, May 2003.
[617] M.A. Brims and B.P. Baxter, *Experience with the OFDA4000 in two mills – comparisons with other instruments*, IWTO Report No. SG 01, Evian, France, May 2004.

[618] V.E. Fish, A. Balasingam, J.W. Marler and T.J. Mahar, *Comparison between OFDA4000 and standardised IWTO test methods, Part 2: Diameter characteristics*, IWTO Report No. RWG 02, Biella, Italy, November 2005.

[619] D.R. Caroll, *A proposed draft test method for the measurement of fibre diameter and length characteristics using OFDA4000*, IWTO Report No. SG 02, Biella, Italy, November 2005.

[620] V.E. Fish, A. Balasingam, J.W. Marler and T.J. Mahar, *Comparisons between OFDA4000 and IWTO standardised methods for measurements made on wool tops: Part 1: Diameter characteristics*, IWTO Rep. No. SG 02, Biella, Italy, November 2005.

[621] A. Balasingam, V.E. Fish, T.J. Mahar and J.W. Marler, *Comparisons between OFDA4000 and IWTO standardised methods for measurements made on wool tops: Part 2: Length characteristics*, IWTO Report No. SG 03, Biella, Italy, November 2005.

[622] E. Gee, *Staple strength as a measure of the soundness of wool, a preliminary note*, SAWTRI Technical Report No. 429, Port Elizabeth, South Africa, 1978.

[623] L.A. Holt, Wool Technol. Sheep Breed. 44(1) (1996) pp. 73–75.

[624] D. Stevens and D.W. Crowe, *Style and processing effects*, in *Proceedings of the Woolspec. '94*, CSIRO Division of Wool Technology, Sydney, Australia, 1994, pp. E1–E12.

[625] E. Gee, SAWTRI Technical Report No. 361, Port Elizabeth, South Africa, 1977.

[626] R.C. Palmer, J. Text. Inst. 40 (1949) p. T623.

[627] C.E. Gore, C.S.P Lee and G.K. van Haaften, Int. Symp. ICC – Congress Centre, 187, Ghent, March 1985.

[628] R.L. Bratt, *A note on some observations on Bradford drawing and spinning*, in *Proceedings of the 3rd Int. Wool Text. Res. Conference*, 4, Cirtel, Paris, 1965, pp. 205–211.

[629] D.W.F. Turpie, M.A. Strydom and E. Gee, *Processing losses during topmaking Part 2: The effect of certain fibre properties and the degree of fribbiness of the raw wool*, SAWTRI Technical Report No. 496, Port Elizabeth, South Africa, 1982.

[630] M.A. Chaudri and K.J. Whiteley, Text. Res. J. 40 (1970) pp. 775–779.

[631] J.F. Sinclair and G.F. Wood, J. Text. Inst. 56 (1965) pp. T274–T279.

[632] W.J. Onions, W. Oxenham and P.A. Townhill, J. Text. Inst. 68 (1977) pp. 370–383.

[633] H.W. Holdsman, Text. Res. J. 39 (1969) pp. 228–233.

[634] A. Samson and M.A. White, Text. Res. J. 40 (1970) pp. 803–809.

[635] S. Yang, N.G. Blenman and P.R. Lamb, *Proceedings of the 3rd Asian Text. Conference*, Hong Kong, 1995.

[636] S. Yang, M. De Ravin, P.R. Lamb and N.G. Blenman, *Wool fibre bundle strength measurement with Sirolan-Tensor*, in *Proceedings of the Top-Tech 96 Symp.*, CSIRO Division of Wool Technology, Geelong, Australia, 1996, pp. 293–304.

[637] Y. Zhang and X. Wang, Wool Technol. Sheep Breed. 49(3) (2001) pp. 212–221.

[638] S. Yang, J.A. Schütz and P.R. Lamb, *Advantages from an improved instrument for the measurement of fibre bundle strength in tops*, IWTO Report No. 6, Boston, USA, May 1997.

[639] C.E. Gore, C.S.P. Lee and G.K. Van Haaften, *Measurement of the physical properties of wool fibre and their relevance in subsequent processing performance*, in *Proceedings of the Int. Symp. 'Western-European Textiles Tomorrow'*, Ghent, Belgium, 1990, pp. 187–209.

[640] S. Yang, *Effect of top dyeing on wool fibre tensile properties and yarn performance*, CSIRO Report, Geelong, Australia, 1999.

[641] J.H. Dittrich, *The effect of top dyeing on spinning performance and yarn properties*, in *Proceedings of the 7th Int. Wool Text. Res. Conference*, II, Tokyo, Japan, 1985, pp. 339–348.

[642] H.J. Henning, Mell Textilber. 52 (1971) pp. 375–380.

[643] J.H. Dittrich, H-J. Henning, *Praxisversuche mit oxidativen filzfrei-ausrüstungen bei gefärbtem material*, Chemiefasern/Textil-Ind. 22/74 (1972) pp. 1124, 23/75 (1973) pp. 49 and 140, Textilveredlung, 9 (1974) pp. 227.

[644] G. Nitschke, Textil-Praxis Int. 28 (1973) pp. 347–352.

[645] L. Hunter, *The effect of fibre properties on spinning performance and yarn properties for after chrome dyed tops*, SAWTRI Technical Report No. 483, Port Elizabeth, South Africa, 1981.

[646] L. Hunter and S. Smuts, *The resistance to compression and bundle tenacity of South African wool tops*, SAWTRI Technical Report No. 409, 1978.

[647] R. Jerke, H. Kleiker and E. Finnimore, *Staining test for prediction of wool top performance*, IWTO Report No. 11, Paris, France, January 1987.

[648] V.D. Burgmann, Text. Res. J. 29 (1959) pp. 901–906.

[649] N.F. Roberts, J.F.P. James and V.D. Burgmann, J. Text. Inst. 51 (1960) pp. T935–T952.
[650] D.F.G. Orwin, J.L. Woods and R.G. Gourdie, *Cortical cell type and wool strength*, in *Proceedings of the 7th Int. Wool Text. Res. Conference*, I, Tokyo, Japan, 1985, pp. 194–203.
[651] S. Smuts, L. Hunter and H.L.J. Van Rensburg, *The effect of fibre diameter and crimp on some single fibre tensile properties of wool*, SAWTRI Technical Report No. 485, 1981.
[652] D.F.G. Orwin, J.L. Woods and K.H. Elliott, *Composition of the cortex of sound and tender wools*, in *Proceedings of the 6th Int. Wool Text. Res. Conference*, Vol. 2, Pretoria, South Africa, 1980, pp. 193–205.
[653] J. Lecomte, G. Mazingue and M. Overbèke, Bull. Inst. Text. France 69 (1957) p. 7030.
[654] M. Feughelman and P.J. Reis, Text. Res. J. 37 (1967) pp. 334–336.
[655] J.F.P. James, J. Text. Inst. 54 (1963) pp. T420–T432.
[656] P. Mason, Text. Res. J. 34 (1964) pp. 747–754.
[657] R.A. Rottenbury, M.R. Bow, W.J. Kavanagh and R.N. Caffin, Wool Technol. Sheep Breed. 29(4) (1981) pp. 143–148.
[658] M.L. Bingham, R.M.W. Sumner, H. Hawker and J.M. Fitzgerald, Proc. N. Z. Soc. Anim. Prod. 43 (1983) pp. 73–78.
[659] L. Hunter, J.B. van Wyk, P.J. de Wet, P.D. Grobbelaar, P.S. Pretorius, V.J. de Morris and W. Leeuwner, *The effects of nutritional and lambing stress on wool fibre and processing characteristics*, in *Proceedings of the 8th Int. Wool Text. Res. Conference*, II, Christchurch, February 1990, pp. 145–156 and Wool Technol. Sheep Breed. 38(4) (1990) pp. 89–91.
[660] M.L. Ryder, *Proceedings of the Inst. Text. De France and Text. Inst. Conference*, 18, Paris, 1975.
[661] L. Hunter, W. Leeuwner, S. Smuts and M.A. Strydom, *The correlation between stade strength and single fibre strength for sound and tender wools*, SAWTRI Technical Report No. 514, Port Elizabeth, South Africa, 1983.
[662] J. Grignet, P. Delfosse, P. Godard, L Lenoir (1) J. Grignet, P. Delfosse, P. Godard, L. Lenoir, W.D. Ainsworth and K.P. Russel (2), Wool & Woollens India, 39, (October/December) (1983).
[663] W. Yu, R. Postle and H. Yan, Text. Asia 32(11) (2001) pp. 38–41.
[664] P.R. McMahon and J.B. Speakman, Trans. Faraday Soc. 33 (1937) p. 844.
[665] K.A. Ferguson, Wool Technol. Sheep Breed. 3(1) (1956) p. 65.
[666] N.F. Roberts, Wool Technol. Sheep Breed. 7(2) (1960) pp. 35–38.
[667] A.J. Gordon and H. Engel, Wool Technol. Sheep Breed. 28(4) (1980) pp. 37–39.
[668] L.G. Butler, Wool Technol. Sheep Breed. 42(3) (1994) pp. 213–230.
[669] A.C. Schlink, J. Lea, A.J.M. Ritchie and M. Sanders, Wool Technol. Sheep Breed. 44(2) (1996) pp. 81–82.
[670] C.J. Scrivener and A.L. Vizard, Wool Technol. Sheep Breed. 45(2) (1997) pp. 120–128.
[671] L.F. Story, N. Z. J. Agric. Res. 2 (1959) pp. 1104–1110.
[672] C.M. Oldham and A.D. Peterson, *The influence of date of shearing on the processing performance to top of mini-commercial consignments of merino fleece wools grown in either South Western or Eastern Australia – 1. Systematic deviations from TEAM predicted values*, in *Proceedings of the 10th Int. Wool Text. Res. Conference*, CDRom, Aachen, Germany, 2000, pp. RW-P4,1–RW-P4,7.
[673] L.F. Story and D.A. Ross, N. Z. J. Agric. Res. 2 (1959) pp. 1096–1103.
[674] M. Huson and P. Turner, Wool Technol. Sheep Breed. 49(1) (2001) pp. 62–72.
[675] K.A. Hansford, *Influence of nutrition and reproduction on length, strength and position of break of Merino wool*, PhD Thesis, University of New South Wales, Sydney, 1989.
[676] A.C. Schlink and P.I. Hynd, *Fibres in tender staples*, in *Proceedings of the Aust. Soc. Anim. Prod., 20th Biennial Conference*, University of Western Australia, Western Australia, July 1994, p. 50.
[677] R.G. Gourdie, *The relationship between intrinsic material strength variation and staple strength in a flock of New Zealand Romney ewes*, PhD Thesis, University of Canterbury, Christchurch, 1989.
[678] M.G. Huson, J.B. Bedson, K.F. Ley and P.S. Turner, *Intrinsic fibre strength of tender wools, Quality Management – Greasy Wool to Yarn*, in *Proceedings of the Top-Tech. '96'*, CSIRO Division of Wool Technology, 1996, pp. 369–375.
[679] N.F. Roberts, J.F.P. James and V.D. Burgmann, J. Text. Inst. 51 (1960) pp. T935–T948.
[680] J.F.P. James and L.J. Smith, J. Text. Inst. 60 (1969) pp. 85–94.

[681] G.A. Carnaby and G.R. Burling-Claridge, Text. Res. J. 66 (1996) pp. 90–98.
[682] R.C. Palmer, J. Agric. Sci. 39 (1949) p. 265.
[683] H.R. Marston, Physiol. Rev. 32 (1952) p. 66.
[684] S. Yang and P.R. Lamb, *A literature survey on wool fibre intrinsic strength*, CSIRO Division of Wool Technology, Geelong, Australia, 1995.
[685] A.N. Thompson, A.D. Peterson, P.J. Hynd and A.J.M. Richie, *The failure properties of single wool fibres in relation to staple strength in merino wool*, in *Proceedings of the 9th Int. Wool Text. Res. J.*, 2, Biella, Italy, 1995, pp. 134–142.
[686] D.R. Scobie, R.J. Walls, L.J. Markham, J.L. Woods and A.R. Bray, Aust. J. Agric. Res. 47 (1996) pp. 1203–1212.
[687] R.G. Gourdie, PhD thesis. University of Canterbury, New Zealand, 1989, p. 309.
[688] J. Meybeck and G. Gianola, *Contribution to the study of the breaking stress of wool fibres as a function of their diameter*, in *Proceedings of the 1st Int. Wool Text Res. Conference*, D, Melbourne, Australia, 1955, pp. 134–147.
[689] J.D. Collins and M. Chaikin, J. Text Inst. 59 (1968) pp. 379–400.
[690] S. De Jong, W.J Kavanagh and M.W. Andrews, *Factors contributing to the staple strength of wool*, in *Proceedings of the 7th Int. Wool Text Res. Conference*, II, Tokyo, Japan, 1985, pp. 147–156.
[691] J.L. Smith, P.R. Lamb, I.W. Purvis and W. Humphries, *Staple strength and style grade of lots from a single flock has limited impact on processing performance of superfine hogget wools*, in *Proceedings of the 11th Int. Wool Text Res. Conference*, CDRom, 139YF, Leeds, United Kingdom, 2005.
[692] P.R. Lamb, I.W. Purvis and G.A. Robinson, Asian-Aust. J. Anim. Sci. 23 (2000).
[693] A.D. Peterson, S.G. Gherardi and M.R. Ellis, Aust. J. Agric. Res. 23 (1998) p. 1181.
[694] P. Kenny and M. Chaikin, J. Text. Inst. 50 (1959) pp. T18–T38.
[695] S.M.A. Shah and K.J. Whiteley, J. Text. Inst. 57 (1966) pp. T286–T293.
[696] W.H. Westenberg and J. Westerveld, *La determination de la resistance specifique de la laine vierge et al. terèe*, in *Proceedings of the 3rd Int. Wool Text. Res. Conference*, II, Paris, France, 1965, pp. 11–19.
[697] W. Yu, R. Postle and H. Yan, 6th Asian Text. Conference, Hong Kong, 2001, pp. 1A–2.
[698] J.D. Collins and M. Chaikin, Text. Res. J. 35(9) (1965) pp. 777–787.
[699] G. Mata, N.J. Edwards, J.E. Hocking Edwards and D.G. Masters, Anim. Prod. Aust. 24 (2002) p. 326.
[700] A.E. Dollin, A.C. Schlink and L.N. Jones, *Merino wool fibre fracture morphologies and wool strength*, in *Proceedings of the 9th Int. Wool Text. Res. Conference*, Vol. II, Biella, Italy (1995) pp. 116–123.
[701] J.L. Woods, D.F.G. Orwin and W.G. Nelson, *Variations in breaking stress of Romney wool fibres*, in *Proceedings of the 8th Int. Wool Text. Res. Conference*, I, Christchurch, New Zealand (1990) pp. 557–568.
[702] X. Wang, Y. Zhang, R. Postle and N. Pan, Text. Asia 29(6) (1998) pp. 33–34.
[703] X. Wang, L. Chang and L.J. Wang, J. Text. Inst. 90(Part 1, No. 3) (1999) pp. 456–457.
[704] S. Smuts, L. Hunter and H.L.J. van Rensburg, *Some typical single fibre tensile properties for wools produced in South Africa*, SAWTRI Technical Report No. 482, 1981.
[705] L. Hunter, W. Leeuwner, S. Smuts and M.A. Strydom, *The correlation between staple strength and single fibre strength for sound and tender wools*, SAWTRI Technical Report No. 514, 1983, pp. 1–15.
[706] Y. Zhang and X. Wang, Wool Technol. Sheep Breed. 40(4) (2000) pp. 303–312.
[707] X. Wang, L. Chang and L. Wang, Text. Asia 29(6) (1998) pp. 46–47.
[708] X. Wang, L.L. Chang and L.J. Wang, *Predicting fibre strength variation from fibre diameter variation*, in *Proceedings of the 2nd China Int. Wool Text. Conference*, Xi'an, P.R. China, 1998.
[709] X. Wang, Text. Res. J. 70 (2000) pp. 191–194.
[710] P. Buckenham, K.J. Whiteley, F.F. Scaybrook and E. Prasad, Wool Technol. Sheep Breed. 33(1) (1985) pp. 33–34.
[711] F. Szemes, Wool Technol. Sheep Breed. 29(2) (1981) pp. 85–86.
[712] A.B. Baumann, Wool Technol. Sheep Breed. 29(4) (1981) pp. 165–167.
[713] P. Baxter, Wool Technol. Sheep Breed. 44(1) (1996) pp. 119–137.
[714] A.I. Vizard, C.J. Scrivener and G.A. Anderson, Wool Technol. Sheep Breed. 42(2) (1994) pp. 112–119.

[715]   B.P. Baxter, *Comparative data on length and strength (IWTO-30) determined by ATLAS and Agritest Staple Length and Staple Breaker instruments*, IWTO Raw Wool Group Meeting Appendix 9, Nice, France, 1995.

[716]   Anonymous, *A3 Agritest Staple Breaker Model 2*, IWTO Raw Wool Group Meeting Appendix 6, Boston, USA, May 1997.

[717]   A.L. Vizard, C.J. Scrivener and G.A. Anderson, Wool Technol. Sheep Breed. 42 (1994) pp. 112–119.

[718]   J.P. Kennedy, Wool Technol. Sheep Breed. 31(1) (1983) pp. 20–21.

[719]   R. Thompson, Wool Technol. Sheep Breed. 33(4) (1985) p. 151.

[720]   R.N. Caffin, J. Text. Inst. 71(2) (1980) pp. 65–70.

[721]   R.N. Caffin, J. Text. Inst. 74(2) (1983) pp. 72–82.

[722]   D.W.F. Turpie and J. Cizek, *Further studies involving the SAWTRI length/strength tester*, SAWTRI Technical Report No. 572, 1985 and IWTO Report No. 9, Barcelona, May 1985.

[723]   J. Cizek and D.W.F. Turpie, *The performance and application of the SAWTRI length/strength tester for raw wool*, in *Proceedings of the 7th Int. Wool Text. Res. Conference*, II, Tokyo, 1985, pp. 137–146.

[724]   A. Litherland, D. Paterson and G.A. Newman, *The processing potential of superfine wool produced under varying environmental conditions*, in *Proceedings of the 8th Int. Wool Text. Res. Conference*, II, Christchurch, New Zealand, 1990, pp. 169–179.

[725]   D.A. Ross, N. Z. J. Agric. Res. 3(3) (1960) pp. 503–525.

[726]   R.L. Thompson, M.W. Andrews, A.B. Baumann, M.R. Bow, C.J. Cantrall, P.H. Heuer and P.A. Irvine, J. Text. Inst. 79(1) (1988) pp. 66–78.

[727]   B.P. Baxter, *Staple length and strength: Short staples and second cuts, effects on coefficient of variation of length (CVL)*, IWTO Raw Wool Group Meeting Appendix 7, Dresden, Germany, June 1998.

[728]   R.N. Caffin, J. Text Inst. 74(2) (1983) pp. 72–81.

[729]   W.D. Ainsworth, Wool Technol. Sheep Breed. 35(4) (1987/1988) pp. 216–223.

[730]   P.M. Cranswick, Wool Annu. (1995) pp. 61–63.

[731]   B. Fernandez and J.W. Marler, *Conditioning staples for staple length and staple strength measurements under IWTO-30*, IWTO Report No. RWG 02, Edinburgh, Scotland, May 2007.

[732]   R.C. Couchman, *Proceedings of the Natural Workshop Management for Wool Quality in Mediterranean Environments*, November 1992, pp. 4–18.

[733]   K.D. Atkins, K.A. Coelli, A.E. Casey and S.J. Semple, N.S.W. Agric.Agfact A3.3.33 (1994).

[734]   M. Dolling, C.L. Cotton and K.L. Butler, Wool Technol. Sheep Breed. 42(3) (1994) pp. 268.

[735]   K. Hansford, *Collation and analysis of available information on additional measurement relevant to improved management of sheep*, Report for the SMISA.M. Committee, University of New South Wales, Sydney, Australia.

[736]   D.G. Knowles, *The relationship between New Zealand merino fleece measurement produced at NZPAC Ltd. and their corresponding core test results*, IWTO Report No. RWG 03, Christchurch, New Zealand, April 2000.

[737]   A.R. Lindsay, *Understanding SS 25*, AWTA Newsletter, April 2007.

[738]   A.N. Thompson and P.I. Hynd, Aust. J. Agric. Res. 49 (1998) p. 889.

[739]   T.C. Reid, Wool Technol. Sheep Breed. 46(4) (1998) pp. 318–337.

[740]   J.M. Dyer, S.D. Bringans, G.D. Aitken, N.I. Joyce and W.G. Bryson, Col. Technol. 123 (2007) pp. 54–58.

[741]   T.C. Reid, Proc. N. Z. Soc. Anim. Prod. 53 (1993) p. 315.

[742]   F.J. Aitken, D.J. Cottle, T.C. Reid and B.R. Wilkinson, Aust. J. Agric. Res. 45 (1994) p. 391.

[743]   M.J. Clark and K.J. Whiteley, J. Text. Inst. 69 (1978) pp. 121–130.

[744]   M.J. Clark and K.J. Whitely, Wool Technol. Sheep Breed. 25(1) (1977) pp. 5–9.

[745]   D.A. Ross, G.A. Wickham and K.H. Elliott, *Breeding objectives to improve wool used in carpets*, in *Proceedings of the 1st World Congr. Sheep and Beef Cattle Breed.*, R.B. Barton and W. Smith, eds., Palmerston North, 1980, pp. 37–45.

[746]   E. Wood, Wool Technol. Sheep Breed. 50(2) (2002) pp. 121–132.

[747]   J.W. Marler and M. Samuelsdorff, *Observations on the influence of the colour of wool on its subsequent dyeing performance – Part II. Implications to the presale measurement of colour*, IWTO Report No. 8, Paris, France, January 1987.

[748] K.J. Whiteley, M.J. Clark, S.J. Welsman and J.H. Stanton, J. Text. Inst. 71 (1980) pp. 177–188.

[749] K.J. Whiteley, M.J. Samuelsdorff and J.P. Connell, *Observations on the influence of the colour of wool on its subsequent dyeing performance*, IWTO Report No. 8, Paris, France, 1987.

[750] P.D. Morgan, *Average yellowness in Australian wools, and Research and development,* Australian Wool Testing Authority, Australian Wool Exporter, Wellington, New Zealand, 1996, pp. 9 and 13.

[751] T.J. Mahar and L. Osborne, Wool Technol. Sheep Breed. 44(3) (1996) pp. 219–237.

[752] R.A. Rottenbury, Wool Technol. Sheep Breed. 32(3) Special edition (1984) pp. 190–196.

[753] B. Thompson, Wool Technol. Sheep Breed. 36(4) (1988/1989) pp. 96–103.

[754] S.A.S. Douglas, *Amendments to IWTO-56–99 'Method for the measurement of base and as is colour of raw wool'*, IWTO Raw Wool Group Appendix No. 03, Christchurch, New Zealand, April 2000.

[755] B. Thompson and D.C. Teasdale, Wool Technol. Sheep Breed. 33(4) (1985) pp. 127–128.

[756] B.O. Bateup, Wool Technol. Sheep Breed. 32(3) Special Edition (1984) pp. 167–170.

[757] J.N. Skinner, Wool Technol. Sheep Breed. 5(1) (1958) pp. 139–141.

[758] B.R. Wilkinson, Wool Technol. Sheep Breed. 29(4) (1981) pp. 169–174.

[759] B.R. Wilkinson, Wool Technol. Sheep Breed. 29(4) (1981) pp. 175–177.

[760] B. Thompson and K.J. Whiteley, Wool Technol. Sheep Breed. 33(3) (1985) pp. 91–92.

[761] B.O. Batesch, Wool Technol. Sheep Breed. 32(3) (1984) pp. 163–170.

[762] L.A. Holt, J. Lax and L. Moll, Wool Technol. Sheep Breed. 42(2) (1994) pp. 151–159.

[763] K.L. Ford and D.J. Cottle, Wool Technol. Sheep Breed. 41(2) (1993) pp. 161–172.

[764] Anonymous, Wool Sci. Rev., Part I, 1965, pp. 27–34, Part II, 1965, pp. 12–21 and Part III, 1966, pp. 34–44.

[765] J.C. Brown, J. Soc. Dyers Col. 75 (1959) pp. 11–21.

[766] L.A. Holt, L.N. Jones and I.W. Stapleton, *Interaction between wool weathering and dyeing,* in *Proceedings of the 8th Int. Wool Text Res. Conference,* IV, Christchurch, New Zealand, 1990, pp. 117–126.

[767] H.R. Hadfield and D.R. Lemin, J. Soc. Dyers Col. 77 (1961) pp. 97–106.

[768] J. Schreiber, Australasian Text. 1 (1990) p. 49.

[769] A.E. Henderson, *Yellow discolourations, in 'Growing Better Wool'*, A.H. and A.W. Reed, 1968, pp. 55–62.

[770] R.M.W. Sumner, S. Young and M.P. Upsdell, Proc. N. Z. Soc. Anim. Prod. 63 (2003) p. 158.

[771] M.V. Benavides and P. Maher, Gen. Mol. Biol. 26 (2003) p. 267.

[772] A.P. Mulcock, N. Z. Vet. J. 13 (1965) p. 87.

[773] P. Duffield and D.M. Lewis, Rev. Prog. Color. 15 (1985) pp. 38–51.

[774] E. Wood, Wool Technol. Sheep Breed. 50(2) (2002) pp. 121–132.

[775] B.R. Wilkinson, Wool, The Official Publication of the Massey Wool Association of New Zealand 7(4) (1982) pp. 9–12.

[776] K. Schäfer, Mell English 82 (6/2001) pp. E133–E135.

[777] K.R. Millington, Color Technol. 122 (2006) pp. 169–186.

[778] D. Goddinger, K. Schafer and H. Hocker, Wool Technol. Sheep Breed. 42(1) (1994) pp. 83–89.

[779] J.M. Dyer, S.D. Bringans, G.D. Aitken, N.I. Joyce and W.G. Bryson, *Identification of yellow chromophores in non-scourable yellowed wool,* in *Proceedings of the 11th Int. Wool Text Res. Conference,* CDRom, 101, FWSA, Leeds, United Kingdom, 2005.

[780] W.S. Simpson, *Physics and chemistry of wool yellowing*, WRONZ Report No. R 217, Christchurch, New Zealand, May 1999.

[781] K.R. Millington, Color Technol. 122 (2006) pp. 301–316.

[782] J.A. Maclaren and B. Milligan, *The effects of radiation and electric discharges, Chapter 13 in Wool Science,* The Chemical Reactivity of Wool Fibre, Science Press, Marrickville, New South Wales, Australia, 1981, pp. 219–233.

[783] J.E. Wood, R.R. Sherlock and M.H.G. Munro, Int. J. Sheep Wool Sci. 53(2) (2005) pp. 21–42.

[784] D.J. Brown, Int. J Sheep Wool Sci. 54(3) (2006) pp. 1–16.

[785] J.W. Marler, *Amended version of DTM-56 for raw wool,* IWTO Report No. Appendix 02, Florence, Italy, May 1999.

[786] N. Jackson, Wool Technol. Sheep Breed. 20(2) (1973) p. 5.

[787] M. Corrigan, Wool Technol. Sheep Breed. 32(4) (1984) pp. 176–180 and 189.

[788] B. Thompson, Wool Technol. Sheep Breed. 32(4) (1984) pp. 159–162, 175.

[789] M.T. Pailthorpe, Wool Technol. Sheep Breed. 32(3) Special edition (1984) pp. 152–158.

[790] B. Thompson, Wool Technol. Sheep Breed. 35(3) (1987) pp. 147–155.

[791] B. Thompson, Wool Technol. Sheep Breed. 36(4) (1988) pp. 96–103.

[792] A.R. Lindsay, Wool Technol. Sheep Breed. 44(2) (1996) pp. 138–162.

[793] A.R. Lindsay, Wool Technol. Sheep Breed. 44(3) (1996) pp. 238–245.

[794] G.R. Shepherd, T.J. Mahar and J.W. Marler, *An international round trial comparing certified tile calibrations to standard wool top calibrations*, IWTO Report No. RWG 08, Nice, France, December 1998.

[795] A.R. Lindsay, *The measurement of wool colour in CIE colour space*, IWTO Report No. 3, Cape Town, South Africa, April 1996.

[796] P.R. Greatorex, *Evaluation of alternative preparation procedures for colour measurement on New Zealand wools*, IWTO Report No. 18, Boston, USA, May 1997.

[797] R.G. Stewart and J.L. Hoare, Wool 5(2) (1970) pp. 39–44.

[798] M.J. Hammersley and B. Thompson, *Wool colour measurement*, I. WRONZ Communication No. 27, Christchurch, New Zealand, 1974.

[799] D. Knowles, Wool (2001) pp. 33–35.

[800] D.G. Knowles, *Calculation of appropriate range check values for colour testing in accordance with IWTO-56*, IWTO Report No. RWG 03, Edinburgh, Scotland, May 2007.

[801] M.J. Hammersley, P.E. Townsend, G.F. Grayston and S.L. Ranford, Text. Res. J. 65 (1995) pp. 241–246.

[802] D.W. Crowe and G.R. Shepherd, *An international round trial company certified tile calibrations to standard wool top calibrations for greasy wool*, IWTO Report No. RWG 05, Florence, Italy, May 1999.

[803] K. Qi, C.J. Lupton, F.A. Pfeiffer, D.L. Minikhiem and A.D. Whittaker, Unpublished communication, 1995.

[804] R.A. Foulds, P. Wong and M.W. Andrews, Wool Technol. Sheep Breed. 32(2) (1984) pp. 91–100.

[805] M.R. Fleet, T.J. Mahar and J.A. Turk, Wool Technol. Sheep Breed. 50(4) (2002) pp. 650–656.

[806] M.R. Fleet, Wool Technol. Sheep Breed. 44(4) (1996) pp. 264–280.

[807] D.J. Ramsay and W. Humphries, *Keeping dark and medullated fibre contamination under control*, in *Proceedings of the 11th Int. Wool Text Res. Conference*, CDRom , Leeds, United Kingdom, 2005.

[808] G. Giebel, *Dark and medullated fibres, why are these fibres a concern for a spinner, at the example of Südwolle?* IWTO, Growers, Traders and Early Processors Meeting, Evian, France, May 2004.

[809] R.A. Foulds, *Preparation and dark fibre contamination*, in *Proceedings of the Seminar on Clip Preparation Research*, CSIRO Division of Textile Physics and Australian Wool Corporation, Ryde, Australia, April 1983, pp. 79–91.

[810] M.R. Fleet, Wool Technol. Sheep Breed. 33(1) (1985) pp. 5–13.

[811] M.R. Fleet, I.M. Rogan, R.W. Ponzoni and J.E. Stafford, *The importance of non-fleece pigmentation in Merino sheep*, in *Merino improvement programs in Australia*, B.J. Guirk, ed., Proc. Nat. Symp. Leura., AWC, Melbourne, 1987, pp. 175–180.

[812] M.R. Fleet, J.E. Stafford and C.H.S. Dolling, Anim. Prod. 39 (1984) pp. 311–314.

[813] D. Charlton, Wool Technol. Sheep Breed. 40(3) (1992) pp. 89–92.

[814] R.A. Foulds, C.B. McInnes, G. Blankenburg and H. Philippen, *Factors influencing final inspections of white and dyed woven and knitted apparel fabrics for faults due to dark fibre contamination*, IWTO Report No. 11, Lisbon, Portugal, June 1991.

[815] R.A. Foulds, P. Wong and M.W. Andrews, Wool Technol. Sheep Breed. 32(2) (1984) pp. 91–97.

[816] J.E. Watts, G.D. Merritt, H.W.M. Lunney, N.W. Bennett and J.A. Dennis, Aust. Vet. J. 57 (1981) pp. 372–376.

[817] A. Burbidge and C.B. McInnes, *Dark fibre risk and prediction*, in *Proceedings of the Woolspec. 94, Seminar 'Specification of Australian Wool and its Implication for Marketing and Processing'*, CSIRO Division of Wool Technology and International Wool Secretariat, Sydney, Australia, November 1994, pp. N1–N7.

[818] M.R. Fleet and R.A. Foulds, Wool Technol. Sheep Breed. 36(2&3) (1988) pp. 76–81.

[819] A. Balasingam and T.J. Mahar, *Status report on dark and medullated fibre testing of presale core samples and review of the detection threshold for contaminant medullation*, IWTO T&S Committee Report No. RWG 4, Hobart, Australia, April 2005.

[820] M.R. Fleet, A. Alaya-ay and T.J. Mahar, *Relationship between greasy and processed dark fibre contamination from Damara crossbred lambs in Merino wool*, IWTO Report No. CTF 2, Cairo, Egypt, May 2006.

[821] Anonymous, *Determination of the threshold detection level for contaminant medullated fibres*, (AWTA Ltd., AWI) AWTA Annual Review, Victoria, Australia, 2005–2006, p. 32.

[822] M.R. Fleet, R.A. Foulds, T.J. Mahar and J.A. Turk, Int. J. Sheep Wool Sci. 56(1) (2008) pp. 40–59.

[823] S. Hatcher, R.A. Foulds, R.J. Lightfoot and I.W. Purvis, Aust. J. Exp. Agric. 39 (1999) pp. 519–528.

[824] A. Balasingam, *The definitions of medullation threshold values used by different testing methods to define an objectionable medullated fibre in Merino wool*, A literature review, Australian Wool Testing Authority Ltd., Guildford, Australia, January 2005, pp. 1–32.

[825] P.H. Lemaire, J. Léonard, M. Longrée, J. Muller, H. Galère and J. Grignet, *The Optalyser, a new instrument for the measurement of the cleanliness of wool slivers (Detection of neps, VM, dark fibres etc.)*, in *Proceedings of the 8th Int. Wool Text Res. Conference*, III, Christchurch, New Zealand, 1990, pp. 399–411.

[826] M. Longree. J. Leonard, P. Delfosse, H. Galere and B. Leman, *Cleanliness faults counting by way of the Optalyser: Comparison between two instruments and against visual counting*, IWTO Report No. 10, Cape Town, South Africa, April 1996.

[827] M. Longree and P. Delfosse, Wool Technol. Sheep Breed. 50(1) (2002) pp. 16–26.

[828] J. Lappage and J. Bedford, *The WRONZ Medullameter*, WRONZ Report No. R107, Christchurch, New Zealand, 1983.

[829] M.A. Brims, *Measuring fibre opacity and medullation using OFDA: Theory and experimental results on mohair*, IWTO Report No. 21, New Delhi, India, March 1994.

[830] A. Burbidge, C.B. McInnes and R.A. Foulds, *Development and preliminary evaluation of the CSIRO Dark Fibre Risk Scheme for individual sale lots of Australian wool*, IWTO Report No. 12, Lisbon, Portugal, June 1991.

[831] K.A. Hansford, *AWI Project EC 573 – Managing the risk of dark and/or medullated fibre contamination*, Final Milestone Report No. T04, Guildford, Australia, December 2003.

[832] K.A. Hansford and Australian Industry Working Group, *The development of a quantified risk factor for the presence of dark and/or medullated fibres in Australian Merino wool*, IWTO Report No. 7, Dubrovnik, Croatia, November 2003.

[833] Anonymous, *Dark and medullated fibre risk scheme implementation*, AWTA Ltd. Newsletter, March 2004, p. 6.

[834] K.M.S. Curtis and J.H. Stanton, *Analysis of the curvature of wool offered for auction in Australia*, IWTO Report No. CTF 05, Barcelona, Spain, May 2002.

[835] K.A. Hansford and P.G. Swan, *Australian Wool Innovation 2004 global survey of dark and medullated fibres*, IWTO Report No. CFT 02, Biella, Italy, November 2005.

[836] D.W.F. Turpie and C.H. Steenkamp, *Objective measurement of "Objectionable" medullated fibres in commercial mohair tops using an optical fibre diameter analyser (OFDA) – an introduction study*, Report No. 1 presented to the IWTO Technical Committee, Harrogate, June 1995.

[837] A.P. Maher, B.D. Cassidy and J.A. Lee, *The measurement of medullation percent by volume of wool using an OFDA*, IWTO Report No. 19, Dresden, Germany, June 1998.

[838] Australian Wool Testing Authority Ltd., *CSIRO and AWI Development of an improved test for detection of dark and medullated fibres in presale core samples*, IWTO Report No. RWG 5, Shanghai, China, November 2004. Available at www.awta.com.au/Publications/Research_Papers/Wool_Contamination.htm(2006)

[839] R.A. Foulds, *Dark fibre contamination in wool – Its prediction and ramifications,*in *Proc Seminar, The Technical Feasibility of the Sale of Wool by Description*, H.W.M. Lunney, ed., CSIRO Division of Wool Technology, Sydney, Australia, July 1988, pp. 127–139.

[840] M.R. Fleet, T.J. Mahar and S.I. Denney, Wool Technol. Sheep Breed. 50(2) (2002) pp. 92–101.

[841] R.A. Foulds, P. Wong and M.W. Andrews, *Reference scales for colour levels of dark fibre contaminants in wool products. Part 2: Scales composed of dyed wool fibres*, IWTO Report No. 16, Venice, Italy, May 1982.

[842] K.A. Hansford and the Australian Wool Industry Working Group, *The development of a quantified risk factor for the presence of dark or medullated fibres in Australian Merino wool*, IWTO Report No. RWG 7, Istanbul, Turkey, November 2003. Available at www.awta.com.au/Publications/Research_Papers/Wool_Contamination.htm(2006)

[843] R.A. Rottenbury, A. Burbidge and C.B. McInnes, Wool Technol. Sheep Breed. 43(4) (1995) pp. 328–337.

[844] R.A. Foulds, Wool Technol. Sheep Breed. 32(3) (1984) pp. 163–170.

[845] P. Delfosse and J. Grignet, *The Optalyser: Experiments on a range of tops and comparison of results with the CSIRO Dark Fibre Detector*, IWTO Report No. 21, Nice, France, December 1992.

[846] A. Burbidge, R.A. Rottenbury and R.F. Sunderland, *Dark fibre contamination in wool tops from consignments of Australian wool*, IWTO Report No. 11, Nice, France, November 1993.

[847] P.W. Chudleigh, Text. Res. J. 61 (1991) pp. 568–574.

[848] T. Madeley, R. Postle and T.J. Mahar, Text. Res. J. 68 (1998) pp. 545–552.

[849] J.G. Snyman, Text. Res. J. 33 (1963) pp. 803–809.

[850] J.M. Gillespie, Text. Res. J. 35 (1965) pp. 1128–1129.

[851] T. Nay and H. Johnson, Aust. J. Agric. Res. 18 (1967) p. 833.

[852] T. Nay and P. Williams, Aust. J. Agric. Res. 20 (1969) pp. 169–184.

[853] K.J. Whiteley, Wool Technol. Sheep Breed. 7(2) (1960) pp. 31–33.

[854] N.F. Roberts, Wool Technol. Sheep Breed. 2(2) (1955) pp. 37–40.

[855] M. Matsudaira and S. Kawabata, *Fibre crimp retention during textile processing and its effect on fabric quality*, in *Objective Evaluation of Apparel Fabrics*, R. Postle, S. Kawabata and M. Niwa, eds., The Textile Machinery Society of Japan, Osaka, 1983, pp. 193–202.

[856] L. Wang, X. Liu and X. Wang, *Changes in fibre curvature during the processing of wool and Alpaca fibres and their blends*, in *Proceedings of the Textile Institute 83rd World Conference*, Shanghai, China, May 2004, pp. 449–452.

[857] A.R. Edmunds, *Relaxation & snippet length effects on the 12th series of IH tops*, IWTO Report No. RWG 06, Nice, France, December 1998.

[858] B. Thompson and K.J. Whiteley, J. Text. Inst. 76 (1985) pp. 64–73.

[859] P.G. Swan and T.J. Mahar, *Objective measurement of the bulk properties of Australian carding wools*, in *Proceedings of the 8th Int. Wool Text. Res. Conference*, II, Christchurch, New Zealand (1990) pp. 84–93.

[860] V.E. Fish, T.J. Mahar and B.J. Crook, Wool Text. Sheep Breed. 47(4) (1999) pp. 248–265.

[861] S.L. Ranford and M.J. Hammersley, *A feasibility study examining the core bulk measurement of scoured wool using near infrared analysis*, WRONZ, Christchurch, New Zealand, 1992.

[862] D. Crowe, *A between-instrument comparison of staple crimp frequency measured by Crimp Meters*, IWTO Report No. RWG 02, Hobart, Australia, April 2005.

[863] P. Semmel, *Measurement of staple crimp frequency. Part II: The Crimp Meter-Description*, IWTO Report No. CTF 02, Istanbul, Turkey, November 2003.

[864] P. Semmel, *Measurement of staple crimp frequency. Part III: The Crimp Meter-Initial evaluation*, IWTO Report No. CTF 3, Istanbul, Turkey, November 2003.

[865] L. Tang and D.W. Crowe, *Measurement of staple crimp frequency. Part I: Manual measurements using the Crimp Gauge*, IWTO Report No. CTF 1, Istanbul, Turkey, November 2003.

[866] C.M. van Wyk, Onderstepoort J. Vet. Sci. Anim. Ind. 22 (1946) p. 99.

[867] C.M. van Wyk, J. Text. Inst. 37 (1946) pp. T285–T292.

[868] D.H. Lee and G.A. Carnaby, Text. Res. J. 62(4) (1992) pp. 185–191.

[869] J.P. van der Merwe, *The effect of fibre physical properties on woollen spun yarn and plain knitted fabric properties*, PhD thesis. University of Strathclyde, Glasgow, Scotland, 1985.

[870] C.J. van Luijk, J. Text. Inst. 78 (1987) pp. 45–53.

[871] C.J. van Luijk, *Bulk of core samples*, IWTO Report No. 19, Paris, France, January 1987.

[872] S.L. Ranford, C.J. van Luijk and K.H. Elliott, *Precision of a bulk measurement on New Zealand wool*, IWTO Report No. 5, Rio de Janeiro, Brazil, June 1987.

[873] E.J. Wood, C.J. van Luijk and S.L. Ranford, *Progress on the New Zealand bulk test*, IWTO Report No. 19, Avignon, France, June 1988.

[874] R.E.W. Buchanan, *Objection measurement and specification of New Zealand wools*, Report to Wool Committee, IWTO, Paris, France, December 1988.

[875] J. Bedford, D.A. Ross, G.A. Carnaby and J. Lappage, *WRONZ Fibre Bulkometer, Instruction Manual*, WRONZ, Christchurch, New Zealand, March 1977.

[876] C.J. van Luijk and S.L., Ranford, Wool Technol. Sheep Breed. 34(4) (1986) pp. 141–142.

[877] S.L. Ranford, E.J. Wood, G.R. Burling-Claridge, I.M. Cuthberson and B. Batchelar, *Progress on the New Zealand bulk test*, IWTO Report No. 17, Perth, Australia, May 1989.

[878] M.A. Chaudri, *The influence of single fibre characteristics on the felting and bulk compressional properties of wool*, PhD thesis. University of New South Wales, Kensington, NSW, 1966.

[879] K.J. Whiteley, S.J. Welsman, J.H. Stanton and R. Pattinson, J. Text. Inst. 77 (1986) pp. 1–8.

[880] N. Watson, N. Jackson and K.J. Whiteley, Aust. J. Agric. Res. 28 (1977) pp. 1083–1094.

[881] Australian Standard AS3535–1988 Wool – Method for the measurement of resistance to compression.

[882] M.A. Chaudri and K.J. Whiteley, Text. Res. J. 38 (1968) pp. 897–906.

[883] R.I. Slinger and S. Smuts, *A study of the bulk resistance to compression of wool fibres*, SAWTRI Technical Report No. 89, Port Elizabeth, South Africa, January 1967.

[884] A. Naik, D. Palet and M. Escusa, *Bulk measurement of Spanish wools*, IWTO Report No. 10, Lisbon, Portugal, June 1991.

[885] A.R. Edmunds, Wool Technol. Sheep Breed. 45(3) (1997) pp. 227–234.

[886] P.G. Swan and T.J. Mahar, Wool Technol. Sheep Breed. 48(1) (2000) pp. 54–72.

[887] T.P. Dabbs, H. Van Schie and M. Glass, *The effect of fibre curvature on Laserscan diameter measurement*, IWTO Report No. 02, Nice, France, December 1994.

[888] J.T. Lobb, W. Humphries and G.J. Higgerson, *Fibre curvature measurement using Sirolan Laserscan and OFDA*, IWTO Report No. RWG 07, Nice, France, December 1998.

[889] A.D. Peterson, A. Brims, M.A. Brims and S.G. Gherardi, J. Text. Inst. 89(3) (1998) pp. 441–444.

[890] J.T. Lobb, K.A. Hansford, W. Humphries, P.R. Lamb and S. Yang, *A preliminary study of fibre curvature measurement using Sirolan-Laserscan and its potential applications*, IWTO Report No. 8, Boston, USA, May 1997.

[891] P.R. Greatorex, D.G. Knowles and G.V. Barker, *Evaluation of alternative methods for fibre curvature measurement on New Zealand wools*, IWTO Report No. 18, Nice, France, December 1997.

[892] A.R. Edmunds and R.M.W. Sumner, *Further data on the use of OFDA for estimating raw wool core bulk*, IWTO Report No. 7, Nice, France, December 1996.

[893] R.W.M. Sumner, B.P. Baxter and M.P. Upsdell, *Investigations into the prediction of core bulk using OFDA instruments*, in *Proceedings of the 11th Int. Wool Text. Res. Conference*, CDRom Leeds, United Kingdom, 2005.

[894] B. van Rensburg, Wool Technol. Sheep Breed. 48(3) (2000) pp. 233–252.

[895] V.E. Fish, T.J. Mahar and B.J. Crook, *The influence of preparation techniques on the measurement of fibre curvature*, IWTO Report No. CTF 06, Christchurch, New Zealand, April 2000.

[896] V.E. Fish, Wool Technol. Sheep Breed. 50(4) (2002) pp. 792–797.

[897] V.E. Fish, T.J. Mahar and B.J. Crook, *Fibre curvature morphometry and measurement*, IWTO Report No. CTF 01, Nice, France, November 1999.

[898] V.E. Fish, MRuSc thesis. University of New England, Armidale, Australia, 2002.

[899] P.G. Swan, *Objective measurement of fibre crimp, curvature and the compressional properties of Australian wools*, PhD thesis. University of New South Wales, Kensington, Australia, 1993.

[900] A.R. Edmunds and S.L. Ranford, *Conditioning effects on fibre diameter and curvature of the 12th series of IH tops*, IWTO Report No. RWG 03, Nice, France, November 1999.

[901] A.R. Edmunds, *Relaxation and Snippet length effects on 12th series of IH tops*, IWTO Technology and Standards Committee Report No. RWG 06, Nice, December 1998.

[902] A.R. Edmunds, *A note on the 12th Series of HI-standards*, IWTO Raw Wool Group Appendix 2, Dresden, June 1998.

[903] S.L. Ranford, *Update on the fibre curvature test method*, IWTO Report No. RWG 06, Christchurch, New Zealand, April 2000.

[904] B. Thompson and D.C. Teasdale, *The length of Snippets for FFDA measurement*, IWTO Technical Committee Report No. 6, Ostede, June 1986.

[905] J.W. Marler, *Harmonising fibre curvature measurements between different instruments*, IWTO Report No. SG 03, Edinburgh, Scotland, May 2007.

[906] S.I. Denney, *The effect of Snippet length fibre relaxation on the mean fibre diameter, fibre diameter distribution and fibre curvature measured by LASERSCAN and OFDA*, IWTO Technology and Standards Committee Report No. SG 04, Christchurch, New Zealand, April 2000.

[907] B.P. Baxter, *Proposal for calibrating laboratory preparation systems and instruments to measure curvature on raw wool*, IWTO Report No. RWG 04, Nice, France, November 2002.

[908] V.E. Fish, *The standardisation of the mean fibre curvature using the Series 14 Calibration Tops as a reference*, IWTO Report No. RWG 05, Istanbul/Dubrovnik, November 2003.

[909] V.E. Fish, *The standardisation of the mean fibre curvature using the Series 14 Calibration Tops as a reference – Calibrations derived for mini-cored scoured wool*, IWTO Report No. RWG 03, Evian, France, May 2004.

[910] J.W. Marler, *Recommended changes to IWTO-12 and IWTO-47 to enable the calibration of mean fibre curvature measurements*, IWTO Report No. RWG 01, Beijing, China, April 2008.

[911] P.G. Swan, *Fibre specification and staple structure*, in *Proceedings of the 'Woolspec 94' Seminar 'Specification of Australian Wool and its Implication for Marketing and Processing'*, CSIRO Division of Wool Technology and International Wool Secretariat, Sydney, Australia, 1994, pp. G1–G12.

[912] S. Smuts, L. Hunter and M. Van Rensburg, Wool Technol. Sheep Breed. 49(1) (2001) pp. 53–56.

[913] R.W. Ponzoni, K.S. Jaensch, R.J. Grimson, D.H. Smith and V.M. Ingham, Wool Technol. Sheep Breed. 49(3) (2001) pp. 172–192.

[914] P.G. Swan and T.J. Mahar, *An objective technique for measurement of fibre crimp curvature: Part 1: Metrology*, IWTO Report No. 18, Dresden, Germany, June 1998.

[915] J.C. Greeff and A.C. Schlink, Wool Technol. Sheep Breed. 50(1) (2002) pp. 6–10.

[916] B.J. Crook, R. Nivison and I.W. Purvis, Wool Technol. Sheep Breed. 47(3) (1999) pp. 159–169.

[917] K.J. Whiteley and E. Balasubramanian, J. Appl. Polym. Sci. 18(12) (1974) pp. 3501–3508.

[918] R.A. Rottenbury and L.J. Smith, *The processing consequences of classing fleeces for staple strength*, IWTO Report No. 5, London, 1979.

[919] M.A. Chaudri and K.J. Whiteley, Text. Res. J. 40(4) (1970) pp. 297–303.

[920] S.M.A. Shah and K.J. Whiteley, J. Text. Inst. 62 (1971) pp. 361–374.

[921] K.J. Whiteley, N.R. Watson and O.D. Wilkins, J. Text. Inst. 69 (1978) pp. 41–48.

[922] M.A. Ali, MSc thesis. University of New South Wales, Sydney, 1968.

[923] G.E. Scheepers and R.I. Slinger, *The influence of fibre crimp form on the felting and compressional properties of wool*, SAWTRI Technical Report No. 103, Port Elizabeth, South Africa, 1968.

[924] J.B. Sherman, E. Balasubramaniam and K.J. Whiteley, J. Text. Inst. 59 (1968) pp. 1–9.

[925] H. Zahn and G. Blankenburg, Text. Res. J. 32(12) (1962) pp. 986–997.

[926] O.D. Wilkins, K.J. Whiteley and J.H. Stanton, *Australian superfine wool fibre diameter, staple crimp frequency and resistance to compression characteristics*, IWTO Report No. 6, Christchurch, New Zealand, April 1981.

[927] J.E. Deurden, J. Text. Inst. 20 (1929) pp. T93–T100.

[928] W.R. Lang, J. Text. Inst. 38 (1947) pp. T257–T270.

[929] W.R. Lang, J. Text. Inst. 38 (1947) pp. T241–T256.

[930] K.J. Whiteley and D. Charlton, *'Objective Measurement of Wool in Australia*, Chapter 9, Australian Wool Corporation, Melbourne, Australia, 1973.

[931] K.A. Hansford and W. Humphries, *Preliminary studies on the diameter and crimp measurements of fine wools*, IWTO Report No. 12, Nice, France, December 1997.

[932] S. Smuts, L. Hunter and H.L.J. van Rensburg, *The effect of fibre diameter and crimp on some single fibre tensile properties of wool*, SAWTRI Technical Report No. 485, Port Elizabeth, South Africa, 1981.

[933] K.J. Whiteley, S.J. Welsma and J.H. Stanton, *The resistance to compression of Australian wools*, IWTO Report No. 12, Manaco, June 1980.

[934] L. Hunter, *A summary of SAWTRI's research on wool and wool blends – 1952–1987*, SAWTRI Spec. Publ. WOL 78, Port Elizabeth, South Africa, November 1987.

[935] J.H. Dusenbury and J.H. Wakelin, Text. Res. J. 28(12) (1958) pp. 989–1005.

[936] K.J. Whiteley and D. Charlton, Proc. Obj. Meas. Wool Aust. Paper No. 9, 1973.

[937] R.A. O'Connel and H.P. Lundgren, Text. Res. J. 24(7) (1954) pp. 677–685.

[938] R.M.W. Sumner and M.P. Upsdell, Wool Technol. Sheep Breed. 49(1) (2001) pp. 29–41.

[939] J.L. Dick and R.M.W. Sumner, Proc. Soc. N. Z. Anim. Prod. 56 (1996) p. 314.

[940] J.L. Dick and R.M.W. Sumner, Proc. Soc. N. Z. Anim. Prod. 57 (1997) p. 61.

[941] R.M.W. Sumner, Proc. N. Z. Soc. Anim. Prod. 59 (1999) p. 39.

[942] R.M.W. Sumner and J.L. Dick, Proc. N. Z. Soc. Anim. Prod. 57 (1997) p. 65.

[943] R.M.W. Sumner, R.H. Stobart and D.L. Johnson, Proc. N. Z. Soc. Anim. Prod. 53 (1993) p. 307.

[944] M. Nimbs, L. Hygate and R. Behrendt, Proc. Aust. Soc. Anim. Prod. 22 (1998) p. 396.

[945] L.W. Lockart, J. Aust. Inst. Agric. Sci. 24 (1958) pp. 243–246.

[946] I.W. Purvis, *Much ado about crimp*, CSIRO Finewool Newsletter, Geelong, Australia, 1997.

[947] Y.E. Goldsworthy and W.R. Lang, J. Text. Inst. 45(1954) pp. T755–T773.

[948] M.G. Huson, *The effect of curvature and set on the tensile properties of wool fibres*, in *Proceedings of the 8th Int. Wool Text. Res. Conference*, I, Christchurch, New Zealand, 1990, pp. 599–608.

[949] N. Jackson and R.A. Rottenbury, *Style metrology*, in *Proceedings of the 'Woolspec 94', Seminar on specification of Australian Wool and its Implication for Marketing and Processing*, CSIRO Division of Wool Technology and International Wool Secretariat, Sydney, Australia, November 1995, pp. C1–C14.

[950] G.W. Wall, J. Text Inst. 54 (1963) pp. T79–T87.

[951] G.W. Walls, Objective Measurement Symposium, Australia Paper No. 3, Melbourne, Australia, 1974.

[952] C.R. Winston, Wool Technol. Sheep Breed. 37(1) (1989) pp. 27–32.

[953] D.W. Crowe and D. Stevens, *Style and processing effects*, in *Proceedings of the Woolspec 94, Seminar 'Specification of Australian Wool and its Implication for Marketing and Processing'*, CSIRO Division of Wool Technology, Sydney, Australia, November 1994, pp. F1–F11.

[954] Anonymous, *Doing it with Style*, CSIRO Textile News No. 25, Belmont, Australia, May 1995, p. 1.

[955] C.J. Scrivener, A.L Vizard and K.A. Hansford, *The valuation of superfine wools in relation to their topmaking performance*, IWTO Report No. CTF 2, Florence, Italy, June 1999.

[956] D. Stevens, *Handle: Specification and effects*, in *Proceedings of the Woolspec 94, Seminar 'Specification of Australian Wool and its Implication for Marketing and Processing'*, CSIRO Division of Wool Technology and International Wool Secretariat, Sydney, Australia, November 1994, pp. H1–H10.

[957] J.R. Paynter, Wool Technol. Sheep Breed. 11(1) (1964) pp. 69–76.

[958] C.R. Winston, *Objective measurement and processing consequences of Style and Type*, in *Proceedings of the Seminar Technical Feasibility of Sale of Wool by Description*, CSIRO Division of Wool Technology, Sydney, July 1989, p. 59.

[959] Alumni Web Community, *Australian Wool Sales Statistics 1991–1992 Season*, AWC, Melbourne, 1992.

[960] K. Ford and N. Jackson, *Style and environment*, in *Proceedings of the Woolspec 94, Seminar 'Specification of Australian Wool and its Implication for Marketing and Processing'*, CSIRO Division of Wool Technology and International Wool Secretariat, Sydney, Australia, November 1994, pp. K1–K14.

[961] G.M. Abbot, Wool Technol. Sheep Breed. 27(2) (1979) pp. 29–33.

[962] M. Lipson, J.S. Ellingworth and J.F. Sinclair, Text. Inst. Ind. 8 (1970) pp. 100–102.

[963] D.W. Crowe, G.P Davis, K.J. Whiteley, L.J. Smith, H.Z. Ma and B.D. Zheng, Wool Technol. Sheep Breed. 44(1) (1996) pp. 17–28.

[964] G.H. Higgerson and K.J. Whiteley, *Measurement of crimp characteristics of greasy wool staples*, in *Proceedings of the 7th Int. Wool Text. Res. Conference*, II, Tokyo, Japan, 1985, pp. 157–166.

[965] G.H. Higgerson and K.J. Whiteley, *Image analysis techniques for the measurement of staple shape*, IWTO Report No. 15, Paris, France, 1987.

[966] W.R. Lang and A. Skertchly, J. Text. Inst. 46(6) (1955) pp. T433–T439.

[967] Australian Wool Testing Authority, *Final report on the performance of style instruments*, IWTO Report No. RWG 03, Nice, France, November 2001.

[968] R.D. Wright, D.W. Crowe and R.A. Rottenbury, *The objective measurement of style and type*, in *Proceedings of the 8th Int. Wool Text. Res. Conference*, II, Christchurch, New Zealand, 1990, pp. 58–67.

[969] Australian Wool Testing Authority Ltd., CSIRO and The Woolmark Company, *Final report on the performance of the style instruments*, IWTO Report No. RWG 03, Nice, France, November 2001.

[970] Australian Wool Testing Authority and The Woolmark Company, Wool Technol. Sheep Breed. 50(1) (2002) pp. 76–88.

[971] K.A. Hansford, *The variation in relationships between the objectively measured components of greasy wool style*, IWTO Report No. 23, Dresden, Germany, 1998.

[972] K.A. Hansford, G.L. Higgerson, W. Humphries, T. Tchen, C. Tischler, G.H. Brown and P.D. Morgan, *The variation in the relationships between the objectively measured components of greasy wool style*, IWTO Technical Committee Meeting Report No. 23, Dresden, June 1998.

[973] W. Humphries, *Instrument for the objective measurement of style*, in *Proceedings of the Woolspec 94, Seminar 'Specification of Australian Wool and its Implication for Marketing and Processing'*, CSIRO Division of Wool Technology and International Wool Secretariat, Sydney, Australia, November 1994, pp. D1-C4.

[974] P.J. Kruger, J. Text. Inst. 63 (1972) pp. 179–181.

[975] S.W.P. Cloete and E.F. De Beer, Elsenburg J. No. 2 (1997) pp. 12–14.

[976] S.W.P. Cloete and E.F. De Beer, Elsenburg J. No. 1 (1995) pp. 1–4.

[977] J.J. Olivier and S.W.P. Cloete, *Relationship between type and production traits in South African Merino sheep*, in *Proceedings of the 35th Congress South African Soc. Anim. Sci.*, Nelspruit, South Africa, July 1997, pp. 149–152.

[978] G.A. Robinson, Wool Res. 145 (3495) (1986) pp. 54–55.

[979] R.E. Belin, D.S. Taylor and G.W. Walls, *Some comparisons between noble and rectilinear combing*, in *Proceedings of the 3rd Int. Wool Text. Res. Conference*, 4, Paris, 1965, pp. 71–81.

[980] A. Samson and M. Chaikin, *The measurement of fibre entanglement with particular reference to scoured wool*, in *Proceedings of the 3rd Int. Wool Text. Res. Conference*, IV, Paris, 1965, pp. 11–20.

[981] A. Samson, PhD thesis. University of New South Wales, Sydney, 1968.

[982] A. Samson, J. Text. Inst. 63 (1972) pp. 177–179.

[983] B.O. Bateup and J.R. Chistoe, *Siroscour: Study of technical innovation*, in *Proceedings of the Top-Tech. '96 Symp.*, Geelong, Australia, 1996, pp. 419–431.

[984] B. Harrowfield, *The management of fibre entanglement: Summary*, in *Proceedings of the Top-Tech '96 Symp.*, Geelong, Australia, 1996, pp. 91–92.

[985] P. Delfosse, M. Longree, C. Bernardin, P. Lemaire and B. Paquet, IWTO Technology and Standard. Committee Report No. STG 04, 1998.

[986] R.K. Garnsworthy, R.L. Gully, R.P. Kandiah, P. Kenins, R.J. Mayfield and R.A. Westerman, CSIRO Division of Wool Technology Report No. G64, Geelong, Australia and Australian Text. 8(4) (1988) pp. 26–29.

[987] G.R.S. Naylor, D.G. Phillips, C.J. Veitch, M. Dolling and D.J. Marland, Text. Res. J. 67(4) (1997) pp. 288–295.

[988] G.R.S. Naylor, Wool Technol. Sheep Breed. 40 (3/4) (1992) pp. 14–18.

[989] M. Dolling, D. Marland, G.R.S. Naylor and D.G. Phillips, Wool Technol. Sheep Breed. 40 (6/7) (1992) pp. 69–71.

[990] G.R.S. Naylor and J. Stanton, Wool Technol. Sheep Breed. 45(4) (1997) pp. 243–255.

[991] J.M.A. O'Keefe, PhD thesis. University of New South Wales, Sydney, 2000.

[992] S. Yang, R.L. Thompson, M. de Ravin and N.G. Blenman, *An improved method for the fibre bundle test*, in *Proceedings of the 9th Int. Wool Text. Res. Conference*, Vol. IV, Biella, Italy, 1995, pp. 107–113.

[993] P. Baxter and G. Wallace, *Development of a portable greasy wool yield instrument*, IWTO Report No. CTF 01, Cairo, Egypt, May 2006.

[994] C.M. Bartle, Appl Radiat. Isot. 50 (1999) pp. 859–866.

[995] C.B. Franklin, L. Hunter and P.E. Monale, *Characterisation of raw wool using gamma-rays and fast neutrons*, in *Proceedings of the 10th Int. Wool Text. Res. Conference*, CDRom, Aachen, Germany, 2000.

[996] G.M. Abbott and I. Blanchonette, *Wool packaging contamination and all that stuff*, in *Proceedings of the Top-Tech. '96 Symp.*, Geelong, Australia, 1996, pp. 116–124.

[997] C. Iype and T.R. Wan, Text. Res. J. 68 (1998) pp. 889–899.

[998] G.A. Robinson, M.W. Prins and M.G. Haigh, *The importance of neps in worsted processing*, IWTO Report No. 16, Nice, France, December 1997.

[999] S.L. Ranford, M.W. Ellery and R.J. Walls, *Applications of visible/near infrared analysis in the textile industry*, IWTO Report No. CFT 05, Nice, France, November 2002.

[1000] W.F. McClure and M.J. Hammersley, *Fourier analysis of the near-infrared spectra of wool*, in *Proceedings of the 8th Int. Wool Text. Res. Conference*, II, Christchurch, New Zealand, 1990, pp. 230–240.

[1001] M.J. Hammersley, S.L. Ranford and P.E. Townsend, *Near infrared analysis of wool; the calibration transfer problem*, in *Proceedings of the 8th Int. Wool Text. Res. Conference*, II, Christchurch, New Zealand, 1990, pp. 218–229.

[1002] E. Wood, Wool Technol. Sheep Breed. 51(3) (2003) pp. 272–290.

[1003] *Proceedings of the Top-Tech.'96 Symposium*, CSIRO and IWS, Geelong, Australia, 11–14 November 1996.

[1004] Proc. Seminar. *Specification of Australian wool and its Implications for marketing and processing*, CSIRO Division of Wool Technology and International Wool Secretariat (No. 1994), Sydney, Australia.

[1005] Proc. Objective Measurement Symposium, *The Australian wool corporation*, Melbourne, Australia, March 1974.

[1006] R.C. Couchman, *Summary report on the Australian staple measurement adoption programme (ASMAP)*, IWTO Report No. 4, Nice, France, December 1992.

[1007] B.P. Baxter, *Sample preparation effects on diameter – Notes from a brief 30 years history*, IWTO Sliver Group Appendix 1, Boston, USA, May 1997.

[1008] D.G. Knowles and J.W. Marler, *The relationship between fibre diameter measurements by Airflow, Laserscan and OFDA for Australian and New Zealand wools*, IWTO Report No. RWG 02, Florence, Italy, May 1999.

[1009] J.W. Marler, G.R. Shepherd and R.G. Barry, *A comparison of histograms from OFDA, Projection Microscope and Laserscan for wool tops*, IWTO Report No. SG 01, Florence, Italy, May 1999.

[1010] K.R. Atkinson, Wool Technol. Sheep Breed. 37(1) (1989) pp. 38–43.

[1011] P.J. Reis, Aust. J. Agric. Res. 43 (1992) pp. 1337–1351.

[1012] R.G. Gourdie, D.F.G. Orwin, S. Ranford and D.A. Ross, Aust. J. Agric Res. 43 (1992) pp. 1759–1776.

Taylor & Francis
Taylor & Francis Group

The Journal of the Textile Institute

Published on behalf of the Textile Institute

Increased to 12 issues per year

EDITOR-IN-CHIEF:

D. Buchanan, *North Carolina State University, USA*

The Journal of the Textile Institute welcomes papers concerning research and innovation, reflecting the professional interests of the Textile Institute in science, engineering, economics, management and design related to the textile industry and the use of fibres in consumer and engineering applications. Papers may encompass anything in the range of textile activities, from fibre production through textile processes and machines, to the design, marketing and use of products. Papers may also report fundamental theoretical or experimental investigations, practical or commercial industrial studies and may relate to technical, economic, aesthetic, social or historical aspects of textiles and the textile industry.

Double centenary celebration – in 2010 the journal celebrates 100 years of publication – and the Textile Institute has been established for 100 years!

Textiles: a Global Vision

To sign up for tables of contents, new publications and citation alerting services visit **www.informaworld.com/alerting**

updates
Taylor & Francis Group

Register your email address at **www.tandf.co.uk/journals/eupdates.asp** to receive information on books, journals and other news within your areas of interest.

Powered by
informaworld

For further information, please contact Customer Services at either of the following:
T&F Informa UK Ltd, Sheepen Place, Colchester, Essex, CO3 3LP, UK
Tel: +44 (0) 20 7017 5544 Fax: 44 (0) 20 7017 5198
Email: subscriptions@tandf.co.uk

Taylor & Francis Inc, 325 Chestnut Street, Philadelphia, PA 19106, USA
Tel: +1 800 354 1420 (toll-free calls from within the US)
or +1 215 625 8900 (calls from overseas) Fax: +1 215 625 2940
Email:customerservice@taylorandfrancis.com

View an online sample issue at:
www.tandf.co.uk/journals/tjti

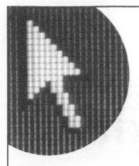

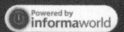

The Textile Institute

Textile Progress

Abstracted and Indexed in:
Compendex
Elsevier Scopus
INSPEC
Textile Technology Index
World Textile Abstracts

Textile Progress is a monograph series that since 1969 has provided critical and comprehensive examination of the origination and application of developments in the international fibre, textile and apparel industry and in its products.

Prospective authors are invited to submit an outline of their proposed contribution for consideration by the Editor-in-Chief to: Professor Xiaoming Tao, Editor-in-Chief, Textile Progress, Institute of Textiles and Clothing, Hong Kong Polytechnic University, Hung Hom, Kowloon, Hong Kong.
Email: tctaoxm@inet.polyu.edu.hk

*Published on behalf of
The Textile Institute by
Taylor and Francis*

Taylor & Francis
Taylor & Francis

Typeset by Aptara, USA

Instructions for Authors

General Style: Contributions suitable for *Textile Progress* should provide a critical and comprehensive examination of the subject matter. For most topics, a manuscript should cover the last five to ten years, using earlier relevant volumes of *Textile Progress* as a starting point. For more specialized or newer material that has not been covered previously in *Textile Progress*, it may be useful to take an earlier starting point.

Contributions should be written in such a way that they allow the non-specialist to understand the principles and applications of the topic. Articles should be broadly based and practically oriented. The appropriate style is that of a monograph, rather than that of a research paper or news article.

Contributions should aim to conform to a printed extent of 48 pp (including tables and illustrations), i.e. approximately 100 manuscript pages. If you feel that the subject requires a longer contribution, please discuss this with the Editor.
Manuscripts should include an appropriate list of references that will allow use as a resource for further study.

Submission of Manuscripts: Prospective authors are invited to submit an outline of their proposed contribution for consideration to:

Professor Xiaoming Tao, Editor-in-Chief,
Textile Progress, Institute of Textiles and Clothing,
Hong Kong Polytechnic University, Hung Hom, Kowloon,
Hong Kong. Email: tctaoxm@polyu.edu.hk.

Except under exceptional circumstances, text, tables and figures should be submitted electronically, along with hardcopy printouts for verification. Tables, figures and text should exist as separate files. *Do not embed either tables or figures in the text*. File names should be self-evident. Files may be prepared using Microsoft Word or LaTex. Please submit a pdf version of your article as well as the LaTex files.

The *Textile Progress* considers all manuscripts on the strict condition that they have been submitted only to the *Textile Progress* on this occasion and that they have not been published already, nor are they under consideration for publication or in press elsewhere. Authors who fail to adhere to this condition will be charged with all costs which *Textile Progress* incurs and their papers will not be published.

Language of Publication: Manuscripts must be written in English. The manuscript will be published in that version of standard English presented by the Author.

Specific Style Guidelines: Informative sub-headings should be used to divide the text. Manuscripts should start with a Table of Contents, in which the major topics are numbered 1., 2., etc., and the major subheadings are numbered 1.1, 1.2 etc. Minor subheadings are numbered 1.1.1,.1.1.2, etc. and 1.1.1.1, 1.1.1.2, etc., if needed. This system should be duplicated in the text of the article. Refer to recent issues of *Textile Progress* for examples.

Tables, figures, and equations are numbered sequentially, starting from 1, without regard to the topic and subhead numbering system. All numbering uses Arabic numerals.

References: References are indicated by bracketed numerals, e.g. [23], [23-26] or [23, 24, 27], starting from 1. In the Reference section, which is the last section of the manuscript, references are listed in consecutive numerical order.

Full reference style details are available on the Textile Progress homepage at www.tandf.co.uk/journals/TTPR.

Figures: Graphs and diagrams should be computer generated when possible and should be supplied in TIF or JPEG format at a resolution of 1200 dpi at their final size. Hard copy graphs and diagrams should be drawn in black ink on good quality paper or supplied as sharp glossy prints. If black-and-white halftones are to be included, the original photographs must be supplied with the manuscript. If originals are not available, electronic files should be supplied at a resolution of 300 dpi at their final size.
If figures are taken from previous publications it is the responsibility of the author to obtain the necessary written permission from the publisher concerned.

A list of captions for figures, including their attribution, if any, must be included. Label each figure on the back with the figure number and the name of the main authors. Illustrations will be reproduced in black and white.

Offprints: Authors will receive 50 free reprints, free online access to their article through our website (www. informaworld.com) and 5 complimentary copies of the issue containing their article. Complimentary reprints are available through Rightslink® and additional reprints can be ordered through Rightslink® when proofs are received. If you have any queries, please contact our reprints department at reprints@tandf.co.uk

Page charges: There are no page charges to individuals or institutions.